TURNING TREES INTO DOLLARS: THE BRITISH COLUMBIA COASTAL LUMBER INDUSTRY, 1858–1913

The coastal forests of British Columbia continue to be a source of controversy, with clashes among corporations, workers, environmentalists, Aboriginal people, and politicians often gaining the attention of the national media. *Turning Trees into Dollars* sheds light on the historical complexities that are the foundation of this ongoing crisis.

This is the first comprehensive account of the formative years of the British Columbian forestry industry. With the concept of market capitalism as his focal point, Hak traces the political, geographical, and technological developments that occurred as small, financially tentative logging operations grew into a major economic force. Hak examines the labour conditions, union activities, use of immigrant labour, and role of the First Nations, as well as the mill industry, tariffs and trade, effects of the Canadian Pacific Railway, industrialization, entrepreneurialism, and conservation movements that compose the rich terrain of B.C. forestry.

This well-documented study provides readers with essential information that has long been missing from public discourse.

GORDON HAK is a professor in the Department of History, Malaspina University-College in Nanaimo.

GORDON HAK

Turning Trees into Dollars
The British Columbia Coastal Lumber Industry, 1858–1913

UNIVERSITY OF TORONTO PRESS
Toronto Buffalo London

BRESCIA COLLEGE
LIBRARY
68719

© University of Toronto Press Incorporated 2000
Toronto Buffalo London

Printed in Canada

ISBN 0-8020-4745-9 (cloth)
ISBN 0-8020-8305-6 (paper)

Printed on acid-free paper

Canadian Cataloguing in Publication Data

Hak, Gordon H. (Gordon Hugh)
Turning trees into dollars : the British Columbia coastal lumber industry, 1858–1913

Includes bibliographical references and index.
ISBN 0-8020-4745-9 (bound) ISBN 0-8020-8305-6 (pbk.)

1. Lumber trade – British Columbia – History – 19th century. I. Title.

HD9764.C33B75 2000 338.4'7674'09711 C99-932894-8

University of Toronto Press acknowledges the financial assistance to its publishing program of the Canada Council for the Arts and the Ontario Arts Council.

This book has been published with the help of a grant from the Humanities and Social Sciences Federation of Canada, using funds provided by the Social Sciences and Humanities Research Council of Canada.

University of Toronto Press acknowledges the financial support for its publishing activities of the Government of Canada through the Book Publishing Industry Development Program (BPIDP).

For my parents,
Hugo and Ingrid

Contents

Acknowledgments

Malaspina University-College provided a release from teaching duties over the 1997–8 academic year, and this gave me the time to complete the project. Archivists and staff at Special Collections, University of British Columbia; Vancouver City Archives; and especially the British Columbia Archives and Records Service in Victoria, where I spent many days, helped uncover the information that supports this study. Three anonymous reviewers carefully read and commented on the entire manuscript. I followed a good number of their suggestions. Gerald Hallowell, and others associated with the University of Toronto Press, effectively guided the manuscript through the various stages to publication. Thanks to all.

TURNING TREES INTO DOLLARS:
THE BRITISH COLUMBIA COASTAL LUMBER INDUSTRY,
1858–1913

Introduction

This study examines the lumber industry in coastal British Columbia in the period 1858 to 1913. The forest industry was not at the forefront of provincial economic development in the nineteenth century: the fur trade, mining, and the fishery generated more economic activity. Lumber was produced, of course, and early observers noted that the forests held great potential for development, but it was not until the first decade of the twentieth century that the forests emerged as a primary industrial focus of the provincial economy.[1] The isolated, financially tentative companies that were operating a few logging camps and sawmills at mid-century expanded into an industry that attracted massive capital investment, employed thousands of workers, put its imprint on large sections of the coastal landscape, and contributed millions of dollars to the provincial treasury.

While the forest industry has long been a significant factor in the history of British Columbia, the attendant historical literature has weaknesses. The best overview of the early history of the industry remains a 1957 MA thesis entitled 'Markets and Capital: A History of the Lumber Industry of British Columbia (1778–1952),' taking the story only up to 1952.[2] G.W. Taylor's *Timber*, a popular history that surveys the industry since the arrival of the Europeans, is informative but lacks scholarly trappings, such as notes, to benefit future researchers.[3]

Specialized studies exist that look at aspects of the industry's history. However, these studies, whether scholarly or popular, tend to take the perspective of one of the three major players in the provincial forest economy, and are concerned primarily with the twentieth century. There are books on prominent companies and lumbermen,[4] studies of government forest management and conservation policy,[5] and accounts of working people in the industry.[6] Few studies include nineteenth-century developments, and the earlier history of government timber policy, conservation, and business remains largely unexplored.[7] Studies of other lumber economies – such as those in Ontario, Quebec, New Brunswick, and the U.S. Midwest – offer helpful lines of inquiry, but the forest types, markets, and timing of the rise of the industries in these areas differ from those in British Columbia. The U.S. Pacific Northwest, with a similar forest cover and geographical setting, offers closer parallels, and histories of this region inform the understanding of the provincial industry traced here.[8] The border, however, was important.

This study fills a gap in the literature by looking at the British Columbia experience before the First World War. It also pulls together a number of interrelated topics: markets, tariffs, lumber companies, timber policy, conservation impulses, logging and sawmilling technology, loggers, and millworkers are discussed here. What gives unity to these diverse topics is that all are a part of one process – the process of turning stands of living trees into cash money. In turn, however, the details of even this process must be seen in a larger context – namely, the spread of capitalism to the west coast of the continent: colonizers, people largely of European descent and driven by profit making, established industries in British Columbia, integrating the region into global trading networks. The patterns and relations of capitalism came to dominate in the region, displacing centuries of First Nations society.[9] With the incursion of international capitalism into British Columbia as the backdrop, this book focuses on the particulars of lumbering.

The general features of the lumbering process were the same in the 1860s and the 1910s. At the physical level, the process involved human beings labouring to transform natural elements into things for human use. In the beginning, there was the forest, a complex natural phenomenon that included trees. In the physical process, human beings converted this living timber into new forms: trees became shingles, doors, sashes, posts, bridge timbers, tongue-and-groove flooring, pickets, rough lumber, finished lumber, and spars. The conversion process involved coping with geography and space. Trees had to be felled, moved, and manipulated, and products had to be transported to final users. This was no simple feat. A production system involving human beings, animals, and machines had to be envisioned, realized, coordinated, and managed.

Physically transforming and transporting timber and wood captures only part of the process; the real magic was in transmuting these things into cash money or its equivalent. Products for use became commodities for exchange in the world of commerce, the world of capitalism. It was the institutions, ideas, and practices of capitalism that allowed a cargo of rough lumber in Burrard Inlet to be exchanged for money that had originated as grain in the Canadian prairies, which in turn could be transformed into machinery from Britain, supplies from California, or luxury goods from France. The successful operation of capitalism depended on a complex array of elements such as systems of credit, business organizations, reliable money and exchange rates, legal systems to protect and encourage commercial trade, protection of private property, and social order. It also depended on the belief of the participants in its efficacy. The lumber industry in coastal British Columbia was enmeshed in this larger context, a system that in the nineteenth century was spreading over the globe.

Distinguishing the physical from the commercial aspects of transforming trees into dollars is, of course, a fiction, for the production system itself was permeated by the dictates of capitalism. Entrepreneurs organized financing, put together companies, and established manufacturing systems in order to produce commodities for markets. The reason for all this activity was to create profits, the fundamental precept of capitalism. So in the establishment and operation of the production system, profit and loss were concerns at all stages. Labour power, machinery, timber, and credit were purchased at the lowest possible cost, and finished commodities were sold at the best price possible. It was not enough to be able to physically produce lumber; it had to be produced in a way that allowed for profit. Production systems that transformed the most timber into profitable market commodities, and did it in the fastest manner, were the most successful. This dynamic engendered competition and ongoing change, whether in establishing new markets, adopting new technologies, or implementing new managerial systems.

The notion of turning trees into dollars had a more literal meaning in 1913 than it did in the 1860s. In the nineteenth century, trees were potential commodities; forests and trees in themselves had no exchange value in the marketplace. Trees had to be transformed into lumber, or at least logs, in order to have value. Indeed, industry people measured standing timber in terms of the lumber that could be produced. They talked of trees in terms of the number of board feet they would eventually produce. One board foot equals a piece of lumber one inch thick and one foot square. A large Douglas fir might yield more than 10,000 board feet, and an especially productive patch of coastal forest might contain more than 100,000 board feet of merchantable timber per acre. Much like the starving wolf in the Saturday-morning cartoons, who, when looking at a live plump chicken, sees it in his thought balloon as cooked, divided into drumsticks and breasts, and steaming on a plate, the nineteenth-century lumberman looked at a tree and envisaged milled lumber in his mind's eye. The standing trees themselves had no use. Only when processed into logs at millside did they have money value. By the end of the period under study, trees were still measured in terms of milled lumber, but now the standing timber itself had commercial value. Changes in global timber supply, markets, and legislation meant that the standing timber in British Columbia forests could be bought and sold. Forests became a commodity that had their own market. Living trees could be turned into dollars, and dollars could be converted into trees in Vancouver, Seattle, and London offices. Lumbermen looked at trees and saw dollars. Market capitalism was gobbling up increasingly large pieces of geography in coastal British Columbia.

As a place, the coastal region was defined by a number of features (see maps on pages 7 and 8).[10] The western border was the Pacific Ocean. The eastern border was formed by the Cascade and Coast Mountains. This zone between the mountains and the ocean contains Vancouver Island and the Lower Mainland, and its climate, topography, and soil created the great temperate rainforests, with their massive cedar and fir trees, the backbone of the coastal forest industry. The northern boundary of the coastal region, defined by geography and forest cover, was roughly at the northern tip of Vancouver Island and went east on the same latitude into the mainland mountains. North of this line the mainland forests changed: Douglas fir and cedar forests gave way to forests of cedar and spruce. Trees also became smaller. Within the mountainous coastal zone, higher elevations also influenced the type of forest cover, and above about 2,000 feet Douglas fir disappeared. For the benefit of Canadians east of the coastal region, J.R. Anderson offered his view of a Douglas fir forest to a convention of foresters in 1901:

> In conclusion let me say to those ... who have not visited the West that they have yet to see a forest in all its magnificence. No other word seems to me to convey a proper idea of a virgin forest of the west. Picture to yourselves thousands of trees, Douglas fir predominating, of prodigious size, so close together that it is with difficulty and often impossible for an animal to go

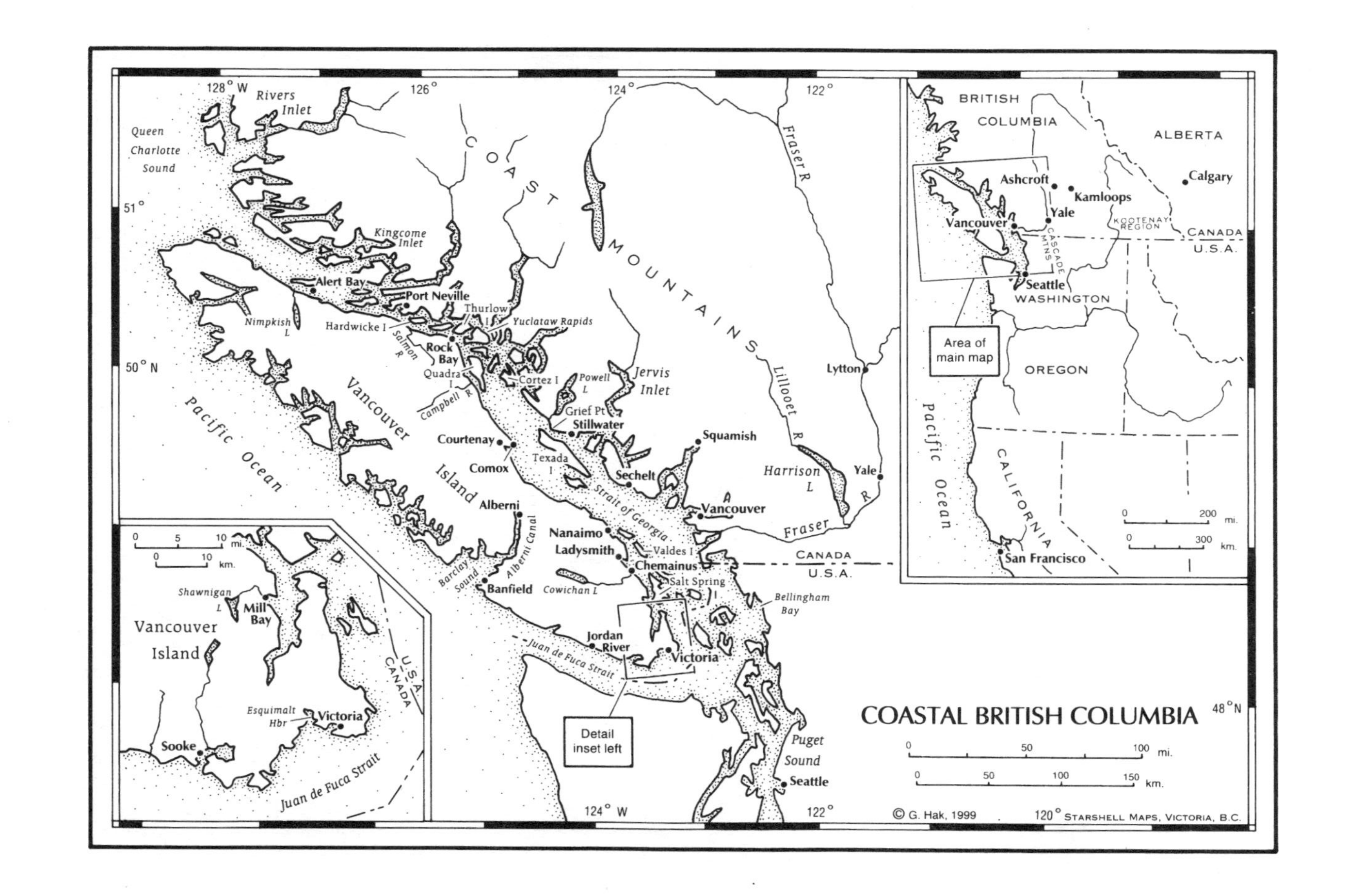
COASTAL BRITISH COLUMBIA
Rivers Inlet
Queen Charlotte Sound
Kingcome Inlet
Alert Bay
Port Neville
Thurlow I
Yuclataw Rapids
Nimpkish L
Hardwicke I
Salmon R
Rock Bay
Quadra I
Campbell R
Cortez I
Powell L
Jervis Inlet
Grief Pt
Stillwater
Courtenay
Comox
Texada I
Sechelt
Squamish
COAST MOUNTAINS
Fraser R
Lillooet R
Lytton
Harrison L
Yale
Vancouver
Fraser R
Vancouver Island
Pacific Ocean
Alberni
Alberni Canal
Barclay Sound
Banfield
Nanaimo
Ladysmith
Chemainus
Valdes I
Strait of Georgia
Salt Spring I
Cowichan L
CANADA
U.S.A.
Bellingham Bay
Jordan River
Victoria
Juan de Fuca Strait
Detail inset left
Puget Sound
Seattle
128° W
126°
124°
122°
51°
50° N
124° W
122°
120°
48° N
0 50 100 mi.
0 50 100 150 km.
© G. Hak, 1999
Starshell Maps, Victoria, B.C.
0 5 10 mi.
0 10 km.
Shawnigan L
Mill Bay
Vancouver Island
U.S.A.
CANADA
Esquimalt Hbr
Victoria
Sooke
Juan de Fuca Strait
BRITISH COLUMBIA
ALBERTA
Calgary
Ashcroft
Kamloops
Yale
Vancouver
KOOTENAY REGION
CASCADE MTNS
CANADA
U.S.A.
Seattle
WASHINGTON
Area of main map
OREGON
Pacific Ocean
CALIFORNIA
San Francisco
0 200 mi.
0 300 km.

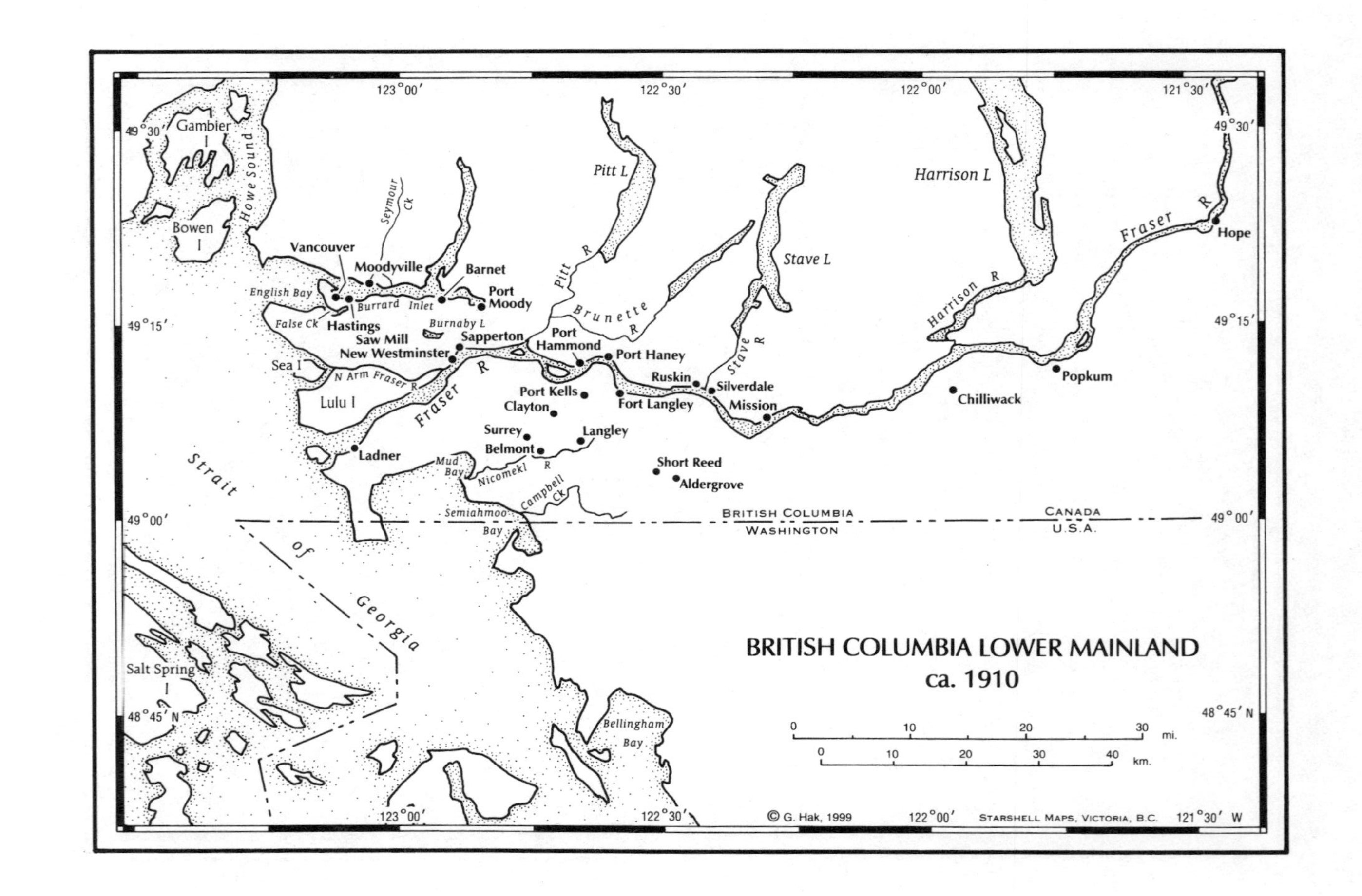
BRITISH COLUMBIA LOWER MAINLAND
ca. 1910
123°00′
122°30′
122°00′
121°30′
49°30′
49°15′
49°00′
48°45′ N
121°30′ W
Gambier I
Bowen I
Howe Sound
Seymour Ck
Vancouver
Moodyville
Barnet
Port Moody
English Bay
Burrard Inlet
False Ck
Hastings
Saw Mill
New Westminster
Burnaby L
Sapperton
Sea I
N Arm Fraser R
Lulu I
Fraser R
Ladner
Pitt L
Pitt R
Brunette R
Port Hammond
Port Haney
Port Kells
Clayton
Fort Langley
Surrey
Belmont
Langley
Mud Bay
Nicomekl R
Campbell Ck
Semiahmoo Bay
Ruskin
Silverdale
Mission
Stave L
Stave R
Harrison L
Harrison R
Chilliwack
Popkum
Fraser R
Hope
Short Reed
Aldergrove
British Columbia
Washington
Canada
U.S.A.
Strait of Georgia
Salt Spring I
Bellingham Bay
0 10 20 30 mi.
0 10 20 30 40 km.
© G. Hak, 1999
Starshell Maps, Victoria, B.C.

between, limbless except the tops through which the rays of sun scarcely penetrate, the ground carpeted with mosses and ferns, and the hush of nature all around you, and you can perhaps form some idea of a forest in British Columbia.[11]

From an economic perspective the boundaries of the region were flexible and changed over time. Large-scale lumbering began on the west coast of Vancouver Island, at Alberni, in the early 1860s, but the focal point of the sawmill industry shifted to Burrard Inlet and New Westminster in the Lower Mainland by the middle of the decade. Over time logging operations pressed eastward up the Fraser River, and up the Stave, Pitt, and Lillooet Rivers. They also spread northward up the mainland coast and up the east coast of Vancouver Island. Logging camps and milling operations became interlinked in an integrated economic unit: logs moved from camps on the northern mainland, in the Fraser Valley, and on eastern Vancouver Island to sawmill centres such as New Westminster, Burrard Inlet, and Chemainus. The ocean water between Vancouver Island and the mainland – the inside passage – was protected from the seas and storms of the open ocean, allowing booms of logs to travel in relative safety from point to point. After the demise of the Alberni operation in the 1860s, the west coast of Vancouver Island, north of the Jordan River, remained outside this trading unit: here it was too dangerous to run log booms on the exposed, open ocean to milling centres such as Victoria or New Westminster.[12]

The southern boundary of the coastal region was established by politics. In 1846 negotiations between Britain and the United States resulted in a border dividing British and American territory. The Oregon Treaty set the boundary at the 49th parallel on the mainland and allotted all of Vancouver Island to the British.[13]

This geopolitical boundary was important to the development of the coastal lumber industry, and its implications help establish 1913 as the closing year of this study. For most years after 1858 the United States maintained a protective tariff against Canadian lumber and other wood products. This tariff helped define the rate and extent of forest exploitation in coastal British Columbia. The tariff not only closed the U.S. market to northern producers, but also slowed the flow of capital investment into the Canadian lumber industry. In 1913 the U.S. government changed its tariff laws and removed the duties on wood, presaging a new era for the British Columbia industry.[14] This study begins in 1858, when a gold rush struck the region, prompting the establishment of the colony of British Columbia. Substantial lumbering began shortly thereafter. In 1858 there were two British colonies in the territory west of the Rockies, British Columbia and Vancouver Island, the latter having been made a colony in 1849. In 1866 these two colonies merged into one, known as British Columbia, and in 1871 the united colony joined Canada as a province.

An interpretative framework that has long underpinned understandings of the British Columbia forest industry is the staples theory. A staple is a natural resource and, in staples theory, analysis of the historical development of resource economies explains that society's particular institutions and way of life. The theory was developed by Harold Innis, a Canadian academic who in the 1920s and 1930s sought to understand Canadian economic development and history. His thought was refined and debated by later scholars.[15]

In a loose version of the theory, the 'staples approach,' observers are content to note common features in resource industries. Production takes place near the resource in isolated, outlying areas; facilities are owned by firms and capitalists located outside the region; capital and labour are imported; the production process is rather simple; and the final goods, raw or semi-processed products, are sold in export markets, where value is added by a more sophisticated labour process. Some or all of these features are common to a number of important Canadian industries, and the 'staples approach' is a well-worn theme in Canadian historiography. One exponent of the approach has claimed that in many respects Canadian history 'is simply the history of fish, furs, lumber, wheat, and so on.'[16] The general emphasis of the staples approach is on markets and the technology needed to get resource products to these markets. W.A. Carrothers's 1938 essay on the British Columbia forest industry is informed by this orientation: Carrothers notes the topographical and climatic features in British Columbia and discusses the technological feats that were necessary to surmount these obstacles and profitably produce lumber.[17]

The more rigorously theoretical 'staples thesis' pushes these ideas further to explain the development of societies based on staples exploitation. In Canada, business and government structures, as well as transportation networks, were created to meet the demands of staple industries, shaping the character of the whole society. Canada was a resource hinterland, dominated by foreign capital, hostage to outside markets, and fundamentally subordinate to foreign metropolitan centres in France, England, and later the United States. In pessimistic versions of staples development, Canada would not emerge as an independent, self-sustaining power unless it threw off this subordinate status, a status that had developed because of the dictates of staples exploitation.[18]

Critics of the staple thesis have decried its simplicity and the lack of empirical evidence. Since the 1970s business studies have shown a more complex picture of Canada's commercial and industrial development than early staples works suggested. Others have argued that the staples focus on markets and trade misses the crucial relationship of capitalist development, the struggle between capital and labour. Moreover, the focus on resource industries in hinterland regions obscures the process of industrialization, a defining feature of modern societies, that was evident in certain areas of Canada in the last half of the nineteenth century.[19]

In British Columbia, an economy that has traditionally relied on resource exploitation, the staples thesis still informs interpretations of economic development.[20] A 1997 collection of scholarly essays on the provincial forest industry is organized around the ideas of Harold Innis, stressing the current relevance of the staples approach.[21] Critics of the British Columbia forest industry in the late twentieth century continue to draw on the pessimistic staples perspective in their analysis. The main features of the forest industry, in their view, is that it is dominated by large, multinational corporations, reliant on export markets, and regulated by a provincial government that shares corporate interests. These features of the forest economy produce an unstable, undiversified, vulnerable provincial economy with a bleak future. Specifically, the critics condemn the export of raw logs, question the flow of profits to international corporations headquartered outside the borders of the province, protest the attitude of these corporations towards the conservation of the resource and the survival of local communities, and decry the failure to further process wood products locally to add more value that supposedly would diversify and stabilize the provincial economy. Patricia Marchak makes the case bluntly: 'Harold Innis would surely have viewed British Columbia between the 1880s and 1970s as the perfect example of the staples economy. Enjoying an embarrassment of resource riches, it failed to develop economic alternatives to export sales of semi-processed products (dimension lumber and pulp).'[22]

This study does not attempt to overthrow the staples approach, which still tells us much about the British Columbia forest industry. Rather, it uses the staples argument to draw out aspects of the early history of the coastal industry, stressing the need to ground notions about resource development empirically and locating staples development in the history of industrial capitalism. Generalizations about staples development based on the structure of the British Columbia forest industry in the middle of the twentieth century do not do justice to the nineteenth-century experience. In the pre-1913 coastal lumber economy, capital initially came from elsewhere to begin commercial lumbering, but the industry was soon largely controlled locally, owned and operated by people who lived in the region. Also, local markets supplemented by the Canadian prairie market, over which British Columbia lumbermen, as Canadians, could claim some political influence, were crucial to the expansion of the industry. External markets, either overseas or in the United States, and external capital, then, do not explain fully the pre-1913 lumber economy. Further, a few large corporations did not dominate the coastal lumber industry by 1913. This was a later development.

Staples theorists also bemoan the unprocessed nature of the final products and the simple, unsophisticated production processes involved. In part, this reflects the desire to emphasize the weakness and undeveloped nature of a staples economy as opposed to a mature industrial, technologically – sophisticated, progressive economy, the ultimate goal of politically active staples critics. This perspective obscures the process of industrialization. If industrialization involves the increased use of steam, and later electrical power, as well as the use of machinery and large-scale production units, a division of labour in the production process, sophisticated coordination of various phases in production, and the existence of a working class, then the lumber economy was clearly industrializing in the late nineteenth century. From an early date sawmills used steam power, with logs passing through a complicated series of machines, manipulated by skilled and semi-skilled workers, before becoming lumber. In the woods, too, steam-powered machinery became increasingly prominent after 1885. Companies were constantly working on new ways to haul heavy logs in an efficient, cheap manner, and to improve cutting technologies. The process of industrialization was not foreign to staples economies.

Staples theory and other theories of industrialization are sometimes seen as being at odds. Staples theory emphasizes market relations to explain economic development. Production for market drives the historical development of capitalism. In the pessimistic version of the staples thesis, exploitation is located in the uneven relationship between the dominant metropolitan centre, which supplies capital and markets, and the subordinate hinterland resource frontier. In conventional Marxist analysis, on the other hand, the relation between capital and wage labour defines capitalism. Exploitation is located in the unequal social and economic relations of class or in the more specific dynamic of the labour process.

This study privileges neither exchange nor production, but takes both seriously in attempting to understand the dynamics of the coastal forest industry. Markets and relations between workers and owners in the production system are both seen as fundamental, with each conditioning the other.[23] The ability of workers in sawmills to make gains in wages and rights was dependent on the condition of the markets in which what they produced was sold. Similarly, the ability of owners to operate profitably in markets was shaped by their ability to dominate, control, and exploit their labour force.

The first three chapters of this study look at the business of lumbering from the perspective of the mill owners. Chapter 1 locates the coastal lumber market in global, local, and continental markets. The importance of various markets changed over time, and political considerations, especially tariff policy, were crucial in defining these markets. Chapter 2 analyses the capitalists and the business structures in the industry. Like lumbering in other parts of North America, at the beginning of the First World War coastal lumber production was not concentrated in a few large firms. Entrepreneurial capital, rather than finance or monopoly capital, operated in a competitive, unstable environment. Chapter 3 discusses how lumber companies coped in this climate of financial instability. As businesses elsewhere did, they sought state help and tried to work together in trade associations to attain a degree of stability.

The next three chapters examine the back end of the industry, the logging and management of timber. Chapter 4 looks at the role of the provincial government in creating a supportive environment for the prosperous prosecution of commercial lumbering, as well as its changing policies on timber tenure. Chapter 5 places the government and industry in a broader political and economic context, uncovering an alternative voice in British Columbia that was critical of government and industry management of the people's forests. Chapter 6 traces the changing structure of the logging industry, especially after 1900, when independent loggers emerged as important economic actors with their own interests, influencing the government's timber policy in the first decade of the twentieth century.

The final three chapters discuss both the mechanical and the human components of the production system. Chapter 7 analyses the material changes in the production process in both the woods and the mills, emphasizing the transition from water and animal power to steam, as well as the sophistication of the constantly changing processes. Chapter 8 looks at loggers, largely single, young white men, who were unable to effect any sustained challenge to capital through organization by the outbreak of the First World War. Chapter 9 discusses millworkers, a workforce that came to be increasingly defined and divided by race.

Studies of business operations, like business itself, must evaluate the meaning of an array of numbers. Quantification is an abstraction devised to simply and meaningfully represent the many aspects of business practices: resource costs, payments to workers, unit production levels, inventory, transportation distances and costs, and profit and loss are all captured in figures. Without access to a complete set of business records from any company, this study perforce pulls together information from diverse and fragmented sources. Moreover, government-generated figures covering the years before 1913 are not as full as they would later be, and their reliability is often suspect. As such, as much as possible I have tried to use figures merely to show general relationships, to give a sense that A was larger than B.

The most important measure is the board foot, which we have already encountered. Mill output, logging-camp production, and the amount of merchantable standing timber in a forest were all calculated in terms of board feet. A sawmill that cut 25,000 board feet a day would produce a stack of lumber roughly 25 feet long, 8 feet high, and 10 feet wide. A tree 6 feet in diameter at the base and 250 feet tall would yield roughly 12,500 board feet of lumber.[24] In a very general sense, I consider sawmills with a capacity of fewer than 20,000 feet per day tiny, those with a capacity of between 20,000 and 50,000 feet small, those with a capacity of between 50,000 and 100,000 feet medium-sized, and those with a capacity of more than 100,000 feet large.

It should also be noted that, in the period under study, people used the British imperial system of measurement, calculating in inches, feet, miles, and acres. I have used this convention, but some might wish to note these metric conversions: 1 inch equals 2.54 centimetres, 1 foot equals 30.48 centimetres, 1 mile is 1.61 kilometres, and 1 acre equals 0.404 hectare.

1

Markets

The fortunes of the pre–First World War coastal lumber economy depended on the demand for wood products. Local mill operators could do only so much to create and cultivate demand for their products, which, for the most part, depended on factors beyond their control. This inability of indigenous lumber companies to control their markets to any great extent was a key feature of the lumber economy. Understanding these markets provides the background to the themes of future chapters.

There were three main markets for coastal lumber. First, from an early date in the industry's development, the Pacific Rim and ports beyond purchased British Columbia lumber products. Export mills serviced this waterborne trade. Second, the local economy, which grew in spurts and starts, was crucial to the fortunes of lumbermen. Population growth and economic activity in Victoria, Nanaimo, Vancouver, and New Westminster, as well as throughout the rural regions of Vancouver Island and the Lower Mainland, stimulated a demand for wood products. The third main market, the Canadian prairies, became significant after the turn of the century.

To become viable, a market had to comprise sufficient economic activity to absorb a substantial quantity of lumber products, but it also had to be accessible to coastal producers. There were two aspects to accessibility. The first one was physical: there had to be the capability for transporting wood products from the mills to their destination. Harbours had to allow ships safe passage, and railway lines had to link producers and distant markets on land. The second aspect was political: government policies had to permit coastal lumber products to be sold. For provincial lumbermen the tariff policies of the Canadian and U.S.

governments were of crucial concern. U.S. tariff policy regulated the importation of goods into the United States, and Canadian policy did the same for Canada. Coastal lumber producers, in the main, found the policies of neither government satisfactory.

Let us begin by looking at the important market that barely existed for coastal producers: the United States. Lack of access to the U.S. market was an enduring theme in the history of the lumber industry between 1858 and 1913. With the exception of the years from 1894 to 1897, the U.S. government imposed a high tariff against imported timber. The United States Tariff Act of 1861 imposed a duty of 30 per cent on 'manufactures of wood' and 20 per cent on 'boards, planks, staves, laths, scantling, spars, hewn and sawn timber.'[1] The 1872 tariff established a rate of $2.00 per thousand board feet against imported rough sawn lumber, and this impost remained in effect until 1894. From 1897 until 1909, under the Dingley Tariff, there was a substantial impost of $2.00 per thousand board feet on rough lumber against imports from Canada. In 1909 changes to the lumber tariff lowered the rate on imports from Canada. The duty on rough sawn lumber, for example, was reduced to $1.25 per thousand board feet. Finally, in 1913 the Underwood Tariff allowed free entry to almost all Canadian wood products.[2] For the period as a whole, depending on economic conditions, the tariff could add 20 per cent or more to the selling price of British Columbia lumber in the United States.[3] There was also ongoing speculation that the tariff would be either lowered or eliminated, and for investors Canadian mills were more valuable if they had access to the large market in the United States.

At an early date, businessmen and government officials recognized the importance of the U.S. tariff to the fortunes of the local lumber industry. In the mid-1850s, Governor James Douglas appealed to the Colonial Office to have Vancouver Island included in the reciprocity treaty concluded in 1854 between the other British North American colonies and the United States, arguing the importance of free access to California markets for Island producers. An address to the first session of the Vancouver Island House of Assembly, on 12 August 1856, stressed the importance of reciprocity. Abolishing the 'fearful imposts,' an import duty of £30 on every hundred pounds' worth of British produce sent to U.S. ports, could be 'hardly overestimated,' said Douglas. With reciprocity, he added, 'the Coal, the Timber, and the productive Fisheries of Vancouver's Island will assume a value before unknown, while every branch of trade will start into activity, and become the means of pouring wealth into the country.'[4]

Douglas's pleas went unheeded.[5] Neither Vancouver Island nor British Columbia participated in the reciprocity treaty. For colonial producers, the high U.S. protective tariff meant that the industry had to survive without access to the larger U.S. market; it also meant that investment in colonial lumbering, whether from Britain or the United States, was slow in coming. Investors recognized the disadvantages faced by the British Columbia producers, and the coastal lumber industry lagged behind developments south of the border. According to one estimate, lumber production in British Columbia in 1869, in millions of board feet, was 25,000, while the state of Washington's was 128,743.[6]

In the 1880s, trade policy and the principles of tariffs were much discussed in the United States, and the tariff was a major issue in the presidential election of 1888.[7] In the late 1880s and early 1890s rumours were rampant that the duty against lumber imported into the United States was about to be lifted. As a New Westminster newspaper editor wrote in 1890, 'the advent of free trade ... is only a matter of time.'[8] Lumber interests on Puget Sound in the United States fought the proposed tariff changes: 'it is impossible to understand how a government can give our markets to our northern neighbors, when the effect must be the wiping out of the present invested interests on the American side.'[9] The *Puget Sound Lumberman* predicted that with free trade British Columbia cargo mills would threaten Washington and Oregon mills in the California trade, and perhaps even compete in the Eastern states.[10] The U.S. lumber lobby was unsuccessful, and the Wilson–Gorman Tariff in 1894 allowed foreign lumber into the United States duty-free. In the midst of a depression, British Columbia lumber exporters took advantage of the situation: in the first half of 1895, forty-six ships took on lumber at provincial export mills and at least half of these were destined for California.[11]

The free access to the U.S. market was not problem-free, however, and coastal lumbermen had to protect their interests. For example, there was a controversy over the U.S. classification of Canadian cedar. The 1894 tariff classified cedar as a hard wood or cabinet wood, and thus subject to a 25 per cent duty. In the fall of 1894, several cargoes of British Columbia cedar were charged duty at U.S. ports, much to the chagrin of British Columbia millmen, who argued that cedar was a soft wood. The Canadians protested, led by J.G. Scott of the Pacific Coast Lumber Company, and the case ended up in the New York state Circuit Court. Scott travelled to Ottawa to enlist the support of the federal government, and then went to New York, armed with samples of sawn British Columbia cedar. Faced with biological and legal evidence, the court accepted the British Columbia view and cedar was accepted duty-free.[12] In 1897, however, the United States reinstituted a general tariff against imported lumber and shingles.

For the most part, then, the United States was not an important market for coastal producers before the First World War. Beyond the short opening of the market from 1894 to 1897, it was only under exceptional circumstances that Canadian wood products penetrated the U.S. market. For example, during the peak years of the California gold rush, the demand for lumber was so great and prices so high that the tariff was not prohibitive. In 1853 the Muir family, operating a sawmill at Sooke on Vancouver Island, sent eighteen cargoes of lumber to San Francisco.[13] Similarly, the San Francisco earthquake and fire in 1906 created a great demand, and the temporary suspension of the duty against Canadian lumber allowed coastal British Columbia mills to contribute to the rebuilding of the California city. The market slackened in late 1907, and by early 1908 mills were closing due to a lumber glut.

After 1900, the shingle sector of the coastal lumber industry had some luck penetrating the U.S. market. Despite a U.S. tariff of $0.30 per thousand against Canadian shingles, coastal shingle mills were more successful in entering the U.S. market than their counterparts in lumber had been. In 1907, 725 million shingles were sold, and 40 per cent of these went to the United States. The quality of British Columbia cedar, the logging-off of cedar trees in the United States, and the development of new shingle machines in British Columbia after 1901, upright shingle sawing machines which produced higher-quality shingles, help explain the ability of Canadian producers to sell in the United States. The U.S. shingle industry struck back in 1909, and when the duty on most lumber products from Canada was reduced, the tariff on shingles increased to $0.50 per thousand.[14] All in all, the coastal lumber industry could not rely on the U.S. market for prosperity.

The waterborne export trade began at an early date. During the 1860s, colonial mills, most notably the Alberni mill, the Moodyville operation, and the Stamp mill on Burrard Inlet, shipped to a wide variety of ports, mainly on the Pacific Rim. Lumber went to Mexico; Sydney and Port Adelaide, Australia; Shanghai and Foochow, China; Valparaiso and Iquique, Chile; Callao, Peru; and the Sandwich Islands. Spar shipments were destined for places such as Batavia, Java; and Cork, Ireland.

To engage in the waterborne trade, the mill had to attract ships, and in the 1860s many shipping companies and captains were concerned about the hazards of Burrard Inlet. Companies constantly emphasized that the Inlet was safe and that facilities were adequate: 'The access to the Mills is up the Gulf of Georgia to English Bay: thence up the Inlet. The anchorage in the bay is safe in all weather. The Inlet is perfectly landlocked, with plenty of water at all times for the largest class vessels. Vessels loading at these Mills will have every facility for loading ... There is a deputy collector at the Mills to enter and clear vessels.'[15] In October 1866, the New Westminster newspaper chronicled the voyage of a ship from San Francisco in the ongoing attempt to overcome prejudices against British Columbia waters:

> The barque *Brazilla*, Capt. McLean, arrived at Messrs. Moody & Co.'s mills, Burrard Inlet, on Sunday last. She came direct from San Francisco to the mills without taking in a pilot, without the aid of steam, and without having touched any port; and, moreover, Captain McLean is a stranger in these waters and had nothing to guide him but the chart and sailing directions. We mention these facts ... in order to expose the wanton malice which has endeavored to create, in San Francisco and elsewhere, the impression that the navigation to Burrard Inlet and New Westminster is difficult and dangerous ...[16]

The export trade had particular demands in delivering lumber cargoes around the world. In 1864, J.A.R. Homer arranged a shipment of lumber to Sydney, Australia, sending his brother-in-law along to supervise.[17] Edward Stamp and S.P. Moody often travelled to San Francisco, the hub of the lumber trade on the west coast, to arrange shipments and secure tonnage. The Moodyville operation had permanent agents, Sproat & Welch, representing them in San Francisco.[18] There were also local agents in the colony that arranged shipments of lumber. In Victoria, John Robertson Stewart was a merchant who participated in the lumber trade in the 1860s, and he had business contacts in Honolulu, Australia, Peru, Chile, and Mexico.[19]

The waterborne export trade was fraught with risk. On 26 February 1864, the *All Serene*, carrying lumber from Vancouver Island to Sydney, Australia, capsized in a hurricane, claiming eight lives and the cargo.[20] In 1867, the *Jeddo* took a cargo of lumber for Moody & Co. to Port Adelaide, Australia. After a journey of sixty-eight days, the ship arrived only to find that the consignees for the shipment had gone bankrupt. The lumber was put up for auction, but due to the simultaneous arrival of several cargoes of Baltic timber, the prices received were very unsatisfactory.[21] Global markets were also volatile; late 1869 and 1870 were particularly bad, and there were long stretches when there were no ships at either the Hastings or the Moodyville wharves.[22]

In the 1870s two export mills on Burrard Inlet serviced the export trade. In the three years from 1876 to 1878, 163 ships took lumber cargoes from these two. As in the 1860s, the Pacific Rim was the primary market. Sixty-three vessels were destined for South American ports, 48 for Australian ports, 25 for Shanghai, and 10 for the Sandwich Islands. A few were bound for Calcutta, Cape of Good Hope, Cork, Liverpool, and London.[23]

In the late 1880s, New Westminster was reactivated as a shipping centre, largely due to the initiative of mill owner John Hendry. New Westminster had been out of the waterborne cargo trade since the mid-1860s. Hendry first began familiarizing overseas customers with lumber products from his New Westminster operation by sending out samples via ships docked in Burrard Inlet. In June 1887, the New Westminster factory sent samples of their doors, windows, sashes, and blinds to Yokohama, Japan, and in November 1887 the plant transported some 50,000 feet of lumber to Burrard Inlet to be included in a shipment to China. In May 1888, they sent 150,000 laths by scow to Vancouver, where they were put on board the steamer to San Francisco.[24] Hendry was also working to entice ships to load directly at New Westminster. He visited San Francisco a number of times, trying to charter vessels, but ship agents and owners 'could not be induced to charter their vessels to load in New Westminster, owing to the supposed dangerous entrance at the mouth of the river, insurance laws and for other reasons.'[25] Due to Hendry's initiative, channel improvements were made, and buoys to guide ships were placed on the river, but in San Francisco in February 1888 he was told that ships would not come until a proper chart of the river was prepared. Hendry secured the support of the New Westminster Board of Trade and prompted Ottawa to begin a survey of the river. Although the chart was not completed, Hendry put up money to cover the risks and chartered a ship. The *Canama*, from Liverpool, arrived in New Westminster on 18 June 1888 and was given a royal reception by excited townspeople: 'the coming of the *Canama* heralds the opening of a new era for New Westminster, and will long be remembered as the first event in the resurrection of the city. Messrs. Hendry & Co. ... will be immortalized in our annals.'[26] On 18 July 1888, the *Canama* left New Westminster with 412,000 feet of lumber bound for Australia; New Westminster was once again a direct participant in the Pacific lumber trade.[27]

There was a big boom in the coastal lumber industry after 1898, but this was not reliant on the waterborne trade. For the most part, the cargo trade was stable: in 1897, some 62,288 million board feet were exported by ship, and in 1906 the total was 74,590 million. There was little variation in the export figures in intervening years.[28] Three mills, Victoria Lumber & Manufacturing, Hastings Saw Mill, and the Moodyville operation, dominated the export trade, accounting for over 90 per cent of exports. They shipped to Britain and Europe, Australia, China and Japan, Africa, and South America.[29] When the prairie market was particularly hot, cargo mills diverted some trade to this region. The Victoria Lumber & Manufacturing Company, located at Chemainus, on the east coast of Vancouver Island, began selling in the prairie market in 1904 because of weak foreign markets. In 1905 the mill exported some 20.3 million feet by ship, and some 5 million feet across the Strait of Georgia and on to the prairies.[30]

Most mills, however, depended on local trade. In almost all populated areas, mills sprang up to serve farmers and townsfolk in the immediate area. Some, in the rural Fraser Valley or the central region of Vancouver Island especially, remained small enterprises supplying the neighbouring community, but others expanded with the growth of their communities.

Small mills had been constructed during the gold rush era at Fort Yale, Lillooet, and Barkerville, producing for the local market.[31] New Westminster, the capital of British Columbia, also supported a number of mills. In 1859 and 1860, Thomas Donahue, a Glaswegian, operated a small mill, and in the first years of the 1860s W.J. Armstrong and J.A.R. Homer had mills. Homer's tiny mill had a daily capacity of 8,000 board feet.[32] On Vancouver Island, the Hudson's Bay Company had constructed a sawmill on Millstream, at the head of Esquimalt harbour, in 1848, and another at Nanaimo in 1854. Officers of the Hudson's Bay Company also established a short-lived sawmill operation at Albert Head, six miles west of Victoria, in 1851. It failed by 1857.[33] By 1853 John Muir and his sons, former coal miners, operated a sawmill at Sooke, and in 1861 William Parsons Sayward, originally from Maine, established a lumbering facility at Mill Bay on the east coast of Vancouver Island, about thirty miles from Victoria. These were small operations that sold lumber in Victoria and Fort Langley on the mainland. In the mid-1860s the Sayward mill was employing some twenty hands in the logging camp and mill.[34]

There was slow but steady growth in the local economy in the 1870s.

The non–First Nations population of British Columbia doubled in the years from 1871 to 1881, to 21,500, the majority living on Vancouver Island and in the Lower Mainland.[35] As well as meeting this local demand, lumbermen positioned themselves for an expected railway connection to the rest of the continent. When British Columbia entered Canada in 1871, the federal government committed itself to building a railway that would link the west coast with the rest of Canada. However, it was not until 1881 that construction began. Industrial development also shaped the fortunes of coastal lumber producers: salmon-canning facilities were established along the Fraser River, with New Westminster acting as the hub for the industry. Salmon canning began on the Fraser in 1867, and ten years later the pack was 55,000 cases; from 1871 to 1883, some twenty-three canneries were constructed on the Fraser River between roughly New Westminster and the ocean.[36]

The beginning of railway construction in the early 1880s spawned excitement in the lumber economy, the first real boom in the history of the coastal lumber industry. The response to the improved conditions would be mimicked in future economic upswings: existing sawmill and logging facilities expanded and newcomers entered the industry, building mills and establishing camps, to take advantage of the buoyant economy. The results were predictable. In 1884, with the end of railway construction in the coastal region and weak markets internationally, the lumber industry was in the doldrums, a situation that extended through 1885. Production was curtailed, companies went bankrupt, and facilities changed hands.

Railway construction gave life to Port Moody. The town, at the head of Burrard Inlet, was declared the terminus of the Canadian Pacific Railway: in early 1882 'the shores of Port Moody presented to view an unbroken line of heavily timbered forests,' but by the end of the summer a railway terminal was constructed and in May 1883 the first ship arrived, bringing steel rails for the Canadian Pacific Railway (CPR).[37] Coastal mills and local entrepreneurs stepped in to service this market. Two New Westminster mills, DeBeck Bros. & Co. and Royal City Planing Mills, established agencies in Port Moody. At Port Moody, John B. Tiffin established the Pioneer Lumber Company, which produced 20,000 feet per day in 1884 as well as shingles, and John Burr and W.J. Armstrong, longtime Fraser Valley residents, established a mill.[38]

The New Westminster mills expanded because of the economic activity associated with railway construction. In March 1882 DeBeck Bros. upgraded their facility, and an observer noted that 'these mills are now ready for the railway "boom."'[39] In the fall of 1882 the mill was busy producing lumber for CPR bridges and other railway works, running two shifts. Further improvements in the winter of 1883–4 increased the capacity of the mill to more than 30,000 feet per twelve-hour shift.[40] The Royal City Planing Mill sawmill was also enlarged to a capacity of 30,000 feet per day in 1883, and a steamer was built to run between the mill and Port Moody.[41]

The completion of the Canadian Pacific Railway in 1885 linked the Far West with the rest of Canada. The first scheduled passenger train arrived in Port Moody, the temporary western terminus of the line, in July 1886. By the fall of 1886, tracks had been constructed to New Westminster, linking that city with points eastward, and in the same year the city of Vancouver was created at the former townsite of Granville on the south shore of Burrard Inlet. Vancouver became the western terminus of the CPR, and the first scheduled train arrived from the east on 23 May 1887.

The railway stimulated economic development in British Columbia, and census population data give a sense of the changing circumstances. In 1881 the population of the province was 49,459; in 1891 it was 98,173; and in 1901 it was 178,657. In 1891 Vancouver Island and the Lower Mainland housed 78.7 per cent of the province's people, including 16,841 in Victoria, 6,678 in New Westminster, and 13,709 in the city of Vancouver.[42] In this expanding local economy, the fortunes of the lumber trade improved measurably. New mills opened to meet the increasing demand. In 1890 there were forty-one sawmills in coastal British Columbia. Five had daily capacities of more than 100,000 feet, and seven had capacities of between 30,000 and 100,000 feet per day. The rest of the operations were small concerns scattered throughout the Fraser Valley and Vancouver Island.[43]

The depression of the 1890s, that lasted from 1892 through 1897, devastated the local lumber economy, but after 1897 the local economy entered another period of frenetic growth. Again, population figures give some insight into the increasing economic activity in British Columbia after 1898. The city of Vancouver's population increased from 27,010 to 100,401 between 1901 and 1911, the area surrounding Vancouver went from 6,069 to 35,890, New Westminster from 6,499 to 13,199, and the Fraser Valley from 12,521 to 29,138. Although less dramatic, there was also population growth on Vancouver Island. According to one estimate the provincial gross domestic product in the first decade of the century was more than double what it was in the 1890s.[44]

Population and economic growth in British Columbia was a boon to the coastal lumber industry after 1898, but this was not the only expanding market. The Canadian prairies finally emerged as an important consumer of coastal wood products. The arrival of the rail link in 1885 led to an increased interest in the continental trade. New Westminster took advantage of the new transportation link at an early date. Royal City Planing Mills shipped lumber over the CPR line during the construction phase, and on 30 September 1886 the company shipped its first lumber to consumers in Calgary and Montreal. John Hendry, the principal owner of Royal City Planing Mills, was instrumental in linking New Westminster mills to eastern markets. To facilitate the trade, he had travelled to CPR headquarters in Montreal in the summer of 1886 to negotiate favourable freight rates for lumber.

The company continued to exploit the rail trade. In January 1888, 350,000 feet of lumber were sent to Port Arthur for the construction of a CPR elevator; in February, carloads of lumber were sent to Regina, and bridge timber was sent to construction sites on the Bow River. From March through June, more bridge timber was destined for the interior of British Columbia, as well as lumber for Medicine Hat and Regina, and for Cobourg, Ontario. The shipment to Cobourg was used in the construction of railway cars. August shipments included 156,000 feet of square timber for the Barney & Smith Car Building Company of Dayton, Ohio: 'these timbers are required to be from 60 to 70 feet in length and without a flaw, as they are used exclusively for car sills and must be perfect.'[45] In a two-week period in early September 1888 Royal City Planing Mills shipped 650,000 feet of lumber east in thirty-four railway cars.[46] In the prairie market itself, growth was slow, reflecting a steady but unspectacular increase in population. The population of Manitoba and the Northwest Territories grew from 118,706 in 1881, to 251,473 in 1891, to 439,641 in 1901.[47]

After the depression of the 1890s, however, the population of the Canadian prairies exploded, and because of the fortunes of the grain trade, farmers had money to spend. In 1911 the three Prairie provinces contained 1.3 million people, more than three times the number for 1901.[48] The demand for lumber in the prairies during boom years was almost insatiable. About 15 million board feet were purchased in 1901 and 1902, 116 million board feet in 1905, and 360 million board feet in 1906.[49]

Despite the growth of the prairie market, it was not safe and secure. While there was little that British Columbia shingle and lumber producers could do about U.S. tariff policy and the U.S. market, they hoped that the Canadian government would be more receptive to their interests, allowing them to manage the prairie trade effectively and profitably. The Canadian federal government, however, turned out to be as unfriendly as the U.S. government.

The problem was simple. As we have seen, in 1894 the government of the United States had dropped duties against Canadian wood products. The Canadian government had responded to the spirit of the agreement by allowing U.S. producers to ship into Canada duty-free. In 1897 the United States reinstituted a tariff against imported lumber and shingles, but the Canadian government did not respond in kind, and lumber remained on the free list. The Laurier government was keen to attract settlers to the prairie West and to maintain the votes of Westerners. Cheap lumber helped meet these economic and political goals.

U.S. mills, especially inland operations, which faced little competition from the Canadians in their home market, shipped surplus lumber north at low prices, dumping it in the Canadian market. In 1905 U.S. mills had approximately 10 per cent of the prairie market, and the Canadians complained that the prairies 'had become the prime dumping ground for low-grade lumber from the Inland Empire.'[50] The British Columbia producers could dominate the provincial market because of distance and freight rates, but to control totally the prairie market they needed the help of Ottawa.

British Columbia producers were furious at the behaviour of the federal government. They had been anticipating an expanding domestic market of their own since the transcontinental railway discussions of the 1870s. However, despite the completion of the Canadian Pacific Railway, settlement in the prairie West did not meet expectations. Then, when growth and expansion became evident in the late 1890s, they were denied reaping all the rewards by the federal government. In the coastal region the British Columbia Lumber and Shingle Manufacturers' Association spearheaded the drive for a fair tariff policy. A few weeks before the lumber producers officially joined the shingle operators in the association, they were already working among themselves to get tariff relief. They met in Vancouver on 2 February 1898 and resolved that 'as long as Canadian lumber going into the United States is taxed, an equal tax should be levied on lumber coming from the States into Canada.'[51]

In early 1898 the lumber and shingle operators enlisted the support of the provincial government. A resolution presented to the legislature stated the manufacturers' case. They noted that the lumber and shingle industries provided employment for a large number of workers in British Columbia, and then outlined the inequities of Canadian tariff legislation. British Columbia producers, they argued, were forced by Canadian customs laws to pay duty on a large portion of supplies and equipment that were imported from the United States. This put them at a disadvantage in competition with U.S. producers, a situation that in the immediate future impeded their ability to capture the market in the Yukon gold rush. Their call for tariff protection from the Canadian government was an attempt to gain equality with U.S. producers. The provincial legislature endorsed the resolution with only three dissenters. There was no response from Ottawa.[52]

The campaign continued. Shingle operator James A. McNair, clearly frustrated by the inactivity of the Canadian government, justified his firm's expansion into the United States by pointing to Canadian tariff policy: 'We do not understand the unfair treatment to the lumber industry in Canada, and we have got tired trying to remedy matters in that way, and are simply going where we can do business and where the lumber industries are protected as well as others.'[53] McNair's shingle mills in Washington state would service the Canadian prairie market as well as the United States. In 1902 the British Columbia branch of the Canadian Manufacturers' Association supported the call for a duty of $2 per thousand board feet on lumber and $0.30 per thousand shingles, sending a petition to the federal Minister of Finance, W.S. Fielding.[54]

In February 1905, a deputation of lumbermen from British Columbia and Ontario met with Prime Minister Laurier in Ottawa. The coastal contingent included T.F. Paterson, John Hendry, and E.H. Heaps. They requested a duty on U.S. lumber, and Paterson provided a list of British Columbia mills that had been forced to close because of the tariff. He argued that the dumping clause should be enforced against lumber from the United States, and if it proved to raise the price of lumber for prairie farmers, too bad: 'the people of the Territories are able to pay the duty. They are, he said, better off than those of British Columbia.'[55]

During Laurier's visit to Vancouver in August 1910, a committee of millmen, including R.H. Alexander, T.F. Paterson, and P.D. Roe, as well as a Kootenay mill owner, again presented their case for tariff protection. Laurier responded by addressing the politics of the tariff and the difficulty of framing a tariff that was suitable to all regions of the country. He had just come from the prairies, where they were clamouring for free trade with the United States. When R.H. Alexander, from the

Hastings mill, lamented that British Columbia mills had been waiting fourteen years for a protective tariff, Laurier replied with a smile and a biblical allusion: 'I would remind Mr. Alexander that Jacob waited and worked fourteen years for his wife.'[56] British Columbia lumbermen had no friend in Wilfrid Laurier. The prairie market remained contested terrain.

By the time of Laurier's 1910 visit, there was a new wrinkle in the tariff debate. Interest groups in the United States were pressing for a trade deal with Canada. By this time, though, British Columbia lumbermen were less keen to have free access to the U.S. market if it meant free U.S. access to the Canadian prairies. Led by newspaper publishers in the United States pushing for free entry of pulpwood from Canada to feed their demand for newsprint, a push for a general free trade or reciprocity agreement with Canada emerged. By 1909 rumours regarding the lowering of duties against Canadian lumber were circulating in British Columbia.

Some British Columbia mill owners felt that open access to the U.S. market would benefit their trade, but most worried that lower production costs in the United States, coupled with the higher transportation costs facing British Columbia mills, would lead to no major market gains south of the border. There was also the opinion that the removal of the tariff against Canadian lumber would spawn a flow of U.S. capital into production facilities in the province, increasing the cut of the forests and shortening the life of the industry. Prominent lumberman and active Liberal supporter John Hendry was against a reciprocity agreement with the United States.[57]

In the end, the negotiated reciprocity treaty was passed by the U.S. government but defeated in Canada in the federal election of 1911. U.S. newspaper publishers, however, were persistent, and a sectoral measure, the Underwood Tariff of 1913, allowed free entry of pulp logs, lumber, and shingles into the United States. For British Columbia coastal producers, this change transformed the industry's marketing strategy over the next decades; in 1929, 55 per cent of British Columbia lumber shipments went to the United States.[58] But because of the depths of the recession in 1913 and 1914, the measure was irrelevant in the short run.

Coastal lumber producers relied on three markets before 1913: the waterborne export trade, the local market, and later the prairie West. After 1858, the ocean trade was strong enough to support two mills on Burrard Inlet, and later a few others elsewhere. Local economic expansion in the 1870s sustained the growth of operations, especially in New

Westminster. With the railway construction boom, the early 1880s saw a short burst of increased activity, as did the first six years after the arrival of the railway. The expanding local market largely explains this growth in the lumber economy of the coastal region, which was curtailed by the depression of the 1890s. The lumber boom after 1898 was fuelled by both the local and the prairie markets. A submission to the 1909 provincial forestry commission gives some sense of the relative importance of these markets, stating that in 1907 the coastal mills sent 70 million feet of lumber by ship, transported 191 million feet by rail to the prairies, and sold 174 million feet locally.[59] Interpretative models, like the staples theory, which stress the importance of external markets in resource-industry development, do not fully capture the pre–First World War situation that existed for lumber producers in British Columbia.

2

Mill Companies

Ownership and capitalization patterns in the coastal milling industry were complicated. Companies entered and left the industry, and saw and shingle mills frequently changed hands, especially in boom times. Ownership instability, then, was one notable feature of the industry. There were also great variations in the size of firms. Some were tiny, one-person operations, while others were substantial corporations that operated a number of milling facilities. This diversity was another feature of the pre-war coastal lumber economy.

Popular histories celebrate the lumber pioneers, focusing on the geographic origins of the entrepreneurs who started large saw and shingle mills.[1] Entrepreneurs and investment came from Britain, the United States, and central Canada and the Maritimes in all decades after 1858, but there were trends. British ownership was especially important through the 1880s, Maritimers made their mark in the 1870s, and central Canadian entrepreneurs began arriving in the 1880s in greater numbers. Notable, too, was the rise of local ownership of the big mill operations in the 1880s.

The scholarly literature on business history and the unfolding of industrial capitalism offers another means to make sense of the industry. Historians, both Marxist and non-Marxist, have outlined a general pattern of business and industrial development that reflects the experience of a number of industries in the nineteenth and early twentieth centuries, industries that grew from a collection of small firms to an oligopoly made up of a few massive corporations.

In early industrial capitalism, industrial firms developed from two sources. In the first scenario, traditional skilled artisans and small producers owned and operated facilities, employing few workers. These were small, personal enterprises. In conducive markets, they expanded their operations by borrowing capital or reinvesting profits; creating factories; adopting industrial techniques; and establishing large, layered business organizations. The second route to industrialization involved merchants – traders – who organized pools of capital and then brought together the ingredients for manufacturing. Merchant businesses dedicated to trade became industrial companies dedicated to production. In both cases the industrial capitalist emerged, replacing the small artisanal producer or the merchant-trader as the key business figure. In an economy or sector of an economy so organized, the owner-entrepreneurs remained close to the production process, managing the day-to-day affairs of the plants. This stage is the era of entrepreneurial, competitive capitalism. Firms and facilities were larger than the small personal enterprises of the artisanal and merchant era, involving more capital, more workers, more technology, and more managers.

Economic expansion generated further changes. Firms grew larger, competition intensified, technologies became more elaborate and expensive, and the need for massive capitalization increased. To cope in this environment, firms pursued a number of strategies. In some economic sectors, companies merged, consolidating operations to overcome the fierce competition. In other sectors, the demand for massive investment attracted financial institutions that had the capital to put together big companies capable of dominating and controlling sectors of the economy. In this stage of capitalist development, companies were massive, operating facilities that were far-flung geographically, and reliant on expensive technologies. Often the corporations owned companies that were involved in a number of different industries. These companies employed great numbers of workers and needed a large, hierarchical management structure to coordinate operations and labour. The entrepreneurial capitalist, who was close to the business and production sides of the operation and physically on the spot, was no longer the key actor. The era of monopoly, finance, or managerial capital, depending on the particulars of the industry and the politics of the writer, had arrived.[2]

It must be emphasized that this schematic overview simplifies what were extremely complex historical processes. There were great variations in the timing and extent of changes, depending on the industry. Still, even in British Columbia resource industries, the general pattern held in the pre–First World War era. Coal mining on Vancouver Island, which began small-scale production at mid-century, was dominated by two large corporations by the 1870s. The hard-rock mining and smelting industry in the Kootenays followed a similar trajectory. Independent prospectors in the 1880s began the rush, and soon larger firms set up mines and smelters. Then, in the first years of the twentieth century, three firms came to control the mining industry in the region. Preeminent was the Consolidated Mining and Smelter Company of Canada, a subsidiary of the Canadian Pacific Railway. The fishing industry also underwent consolidation when a few large canning companies took control of the industry. Up until the late 1880s, the Fraser River canning industry was characterized be a number of small firms, but by 1891 three large firms dominated, and further concentration occurred between 1897 and 1902.[3]

By 1913 the coastal British Columbia lumber industry had not moved beyond the stage of competitive industrial capitalism. The lumber sector remained very decentralized. No firm or group of firms dominated the industry, and while operations got larger and produced more lumber and shingles, small and medium-sized operations continued to operate in a very competitive environment. For example, in 1906, on southern Vancouver Island, Victoria Lumber & Manufacturing was capable of putting out 320,000 board feet per day, the Boyd operation at Cowichan 90,000 feet, the Sayward mill 40,000 feet, and the Shawnigan Lake Lumber Company 30,000 feet. At the same time there was a string of smaller mills in the vicinity, including the Taylor mill (15,000 feet per day); J. Leigh & Sons (15,000 feet); Lemon, Gonnason & Company (10,000); the Toronto and British Columbia Lumber Company's Sidney mill (15,000); the Columbia Mill Company (10,000); and the Ucluelet Mercantile Company (6,000).[4] It was even possible for a few men with little capital to make a go of it. In 1902, George Hugh Pidcock ran a logging operation with a donkey engine and a small sawmill. He logged with the aid of a few helpers, and they could yard twelve logs to the shore with the donkey engine in a day. The steam sawmill produced 1,600 board feet of lumber on a good day.[5]

If the coastal lumber economy did not reach the stage of financial, monopoly, or managerial capitalism by 1913, it had made the transition from merchant and personal-enterprise capitalism to entrepreneurial capitalism. In the 1860s and 1870s, merchant capitalists, and small, personal enterprises, were active in coastal lumbering. Merchant capitalists, with connections to Britain, sustained the big export mills through the lean economic years from the 1860s through to the 1880s. Thereafter the presence of the merchant houses waned. By the 1880s, the dynamic element in the lumber trade was becoming the local market and the continental rail trade, and global connections to markets were less crucial to running a milling operation. Also, by the 1880s there were local capitalists with sufficient capital and expertise to operate the large mills. Locally based ownership took on a more prominent role. Especially successful in making the transition from personal to entrepreneurial enterprises were lumbermen from New Brunswick. One of these transplanted New Brunswickers, John Hendry, became the most notable lumberman of the era.

From the 1880s onward, entrepreneurial, competitive capitalism dominated in the coastal lumber economy, but there were other business forms. The small, personal enterprises did not disappear, and there were a few subsidiary operations, lumber companies with headquarters beyond the borders of British Columbia running mills in the coastal region. In the boom after 1900, there were merger attempts, as well as speculation by interests with little or no connection to lumbering. However, decentralized, competitive capitalism continued to reign. Merchant capital had disappeared from the industry by the early twentieth century, but personal firms and entrepreneurial firms, with the owner-operators resident in British Columbia, had not yet been superseded by more advanced forms of business organization.

The importance of British merchant capital in the colonial economy has long been recognized. Victoria was an economic outpost of San Francisco. British wholesalers lived in San Francisco, maintaining links with businessmen in both colonial British Columbia and Britain.[6] However, the lasting influence of British merchant capital, operating out of San Francisco, in the coastal lumber industry has not been appreciated. San Francisco partners of British merchant houses came to control the two main export mills in the coastal region. In one case, these merchants were reluctant lumbermen, taking on ownership when the mill company who owed them money failed. In the other instance, the merchants were more aggressive in pursuing control of the mill.

For British trading companies, close links to coastal sawmills served a number of purposes. Whether they actually owned a mill or just had exclusive rights to their business, the mills served as customers. Traders could make money selling supplies and machinery to a mill, even if the mill itself was not turning a profit. The mills also provided a cargo to export from British Columbia; importing goods into British Columbia and then having ships leave with no cargo was costly and made it difficult to secure tonnage. Finally, in good market conditions money was made on lumber sales.

Dickson, DeWolf & Company took over the British Columbia and Vancouver Island Spar, Lumber, and Sawmill Company in 1869. The operation had been called Stamp's Mill and now became known as Hastings Saw Mill, or just Hastings mill. Dickson, DeWolf & Company, a British merchant trading company, had had agents in Victoria since at least the mid-1860s, importing goods into the colony and exporting products such as lumber.[7] The history of the Hastings mill goes back to December 1859, when Captain Edward Stamp, an English shipmaster and commission agent, approached the colonial government in Victoria: he was connected with merchants and traders in England who were keen on establishing a sawmill capable of producing 50,000 feet per day, a fishing and fish-curing plant, as well as other industries. The sawmill machinery was already in transit from Britain, as were a number of skilled artisans to run the operation, and Stamp wanted to negotiate a favourable deal with the government of Vancouver Island.[8] Governor Douglas and the colonial administration were eager to attract any industrial development, and a deal was concluded. On 29 June 1860, the schooner *Meg Merrilees* arrived at the head of the Alberni Canal, bringing nine workers to begin construction.[9]

The British investors were associated with the London firms James Thomson & Company and Thomas Bilbe & Company, though Stamp was also an investor.[10] Stamp named the site 'Somass,' and the mill got up steam for the first time on 22 May 1861. In the summer of 1861, the mill's two gang saws were cutting 14,000 feet per day, but there were hopes that they would get a large circular saw that would increase the capacity to 50,000 feet per day.[11] Ships were frequenting the mill, taking on spars and lumber for Australia and England. In 1862 the operation exported 7,490,000 feet of sawn lumber and 990 spars, and this increased to 11,273,000 feet of sawn lumber and 1,300 spars in the following year.[12]

Yet problems were already evident in the summer of 1861. As part of the arrangement with the government, the company had been granted access to some 15,000 acres of timberland. By the summer of 1861 they had taken up about 6,000 acres, but according to Stamp they were not interested in taking up more in the original grant. Instead, Stamp wanted 2,000 acres north of Barclay Sound on the coast.[13] It was becoming clear that the timber supply in close proximity to Somass was inadequate.

In January 1863, the firm was reorganized as Anderson & Co., associated with Anderson, Thomson & Co. in London, and Captain Stamp was no longer involved. Gilbert Malcolm Sproat, who had worked in the London office of James Thomson & Company and arrived at Somass in 1861, became the mill manager. Production in 1863 topped 1862 levels, but the new manager was becoming concerned about the future of the operation. With the mill consuming between 15 and 20 million feet of timber per year, Sproat anticipated that the operation would soon need access to more timber.[14]

By late 1864 the lack of accessible timber was critical. In a letter to the Colonial Secretary, Sproat outlined the plight of the operation. Stamp, he noted, had chosen a poor site, for there were sufficient trees in the immediate vicinity to sustain the milling operation for only about a year and a half. Only by constructing a dam were they able to run logs from a timber stand deeper in the interior. This source was now exhausted. The area was 'totally unsuitable for large Sawmills ... owing to the broken character of the country and the smallness and shallowness of the streams.' The operation was not yielding a profit and continued only 'from an unwillingness to wind it up until forced ...'[15] A few months later the sawmill closed for good.[16]

While Anderson & Co. were content merely to hold their property on Vancouver Island, Stamp was still keen to produce lumber. In 1863 Stamp employed six men to travel the east coast of Vancouver Island and the mainland coast in search of a site suitable for spar-timber and sawmill operations. He initially planned to establish at Port Neville, using Hornsby's Island as a farming centre to provision his operation,[17] but in 1865 he decided to locate at Burrard Inlet. A major reason that he chose Burrard Inlet was that from there he could get logs from the Fraser River area and Howe Sound, as well as from Burrard Inlet.[18] His experience on the Alberni Canal had taught him the necessity of having ready access to a large, secure supply of timber. In April 1865, the company was incorporated as the British Columbia and Vancouver Island Spar, Lumber, and Sawmill Company, but the operation was commonly referred to as Stamp's Mill.

A deal was concluded with the colonial government and construction on the mill began. As construction proceeded, Stamp operated a spar camp. Despite unforeseen obstacles, which forced Stamp to move to a different mill site on Burrard Inlet, and the loss of a portion of mill machinery on the voyage from England, in June 1867 the steam-powered mill began cutting lumber.[19] The spars and lumber produced by the operation were destined for the waterborne export trade.

By 1869, Stamp had fallen out with his British backers, and the two parties were in court, settling their differences. The heavy losses sustained by the sawmill forced the company into liquidation in July 1869. The company was indebted to Dickson, DeWolf & Company, their San Francisco agents, to the extent of about $35,000. Edward Davis Heatley, a British subject, was the associate in San Francisco, and George Campbell was the associate in London. They traded under the name of Dickson, DeWolf & Company.[20] Rather than lose their $35,000, they purchased the mill operation for $14,000 in July 1870 as a strategy to recoup some of their capital. The Hastings mill became a node in the international trading network of a British-based merchant house that was interconnected by a series of associations and partnerships.[21]

Dickson, DeWolf & Company owned the Hastings sawmill throughout the 1870s, leaving the management of the mill to Captain J.A. Raymur. As well as servicing the export trade, the operation also established a mill agent in Victoria and a lumberyard in Nanaimo. In 1874 the mill's orders were sufficient to prompt the operation of two shifts, and with the extra crew the mill ran night and day.[22] In the early 1880s, Hastings Saw Mill had a capacity of 75,000 feet and, through its primary owner, E.D. Heatley of San Francisco, remained part of a global British trading network. E.D. Heatley was now associated with Edwin Wotton and Ernest Lafont Heatley of London in Dickson, DeWolf & Company.[23]

The coming of the Canadian Pacific Railway (CPR) made the Hastings mill site a valuable commodity, and the British merchant Heatley sold the land to British Columbia businessmen, mainly from Victoria, anticipating the extension of the CPR line from the Port Moody terminus to a more westerly point on Burrard Inlet. The Hastings sawmill was situated on a tract of land on the south shore of Burrard Inlet that the mill company owned outright. In the early 1880s, the CPR was keen to extend its line, and many felt that the only way to do this was through the Hastings land. Another variable was that a twenty-one-year timber lease adjacent to the mill, negotiated in 1865 and held by the company, was due to expire in 1886, and the company wanted the lease extended as all the timber had not been logged. The CPR wanted the land for a railway right-of-way, and negotiations involving the company, the railway, and the provincial government began in August 1884. The sawmill company eventually relinquished two timber leases on the south shore of English Bay and False Creek in return for timber on a nearby tract, as well as the rights to log the uncut timber.[24]

Recognizing the potential of the situation emerging in the 1880s, a group of British Columbia businessmen began manoeuvring to take advantage of the possibilities. Victoria businessman Edgar Crow Baker was elected to the federal House of Commons to represent Victoria in 1882. Baker's wife's uncle was Captain James Arnold Raymur, the manager of Hastings Saw Mill, and in the 1870s Baker had served for a time as the Victoria business agent for the firm. In Ottawa, Baker was acquainted with a host of railway people, including A.W. Ross. Ross was the Member of Parliament from Winnipeg, but he had been employed by the CPR as their real estate agent and adviser in the Burrard Inlet region. By 1884, if not sooner, Baker and Ross were convinced that the CPR line was going to be extended beyond Port Moody, and they recognized the business opportunity.[25] Baker, Ross, and other Victoria business people began buying land on English Bay, False Creek, and Coal Harbour. In July 1884, Ross proposed that they buy the Hastings Mill site.[26]

A syndicate was formed. The Coal Harbour Land Syndicate, which was capitalized at $350,000, made up of twenty shares at $17,500 each, was dominated by prominent Victoria businessmen.[27] In the summer of 1884, the syndicate purchased the 304-acre mill site from E.D. Heatley of San Francisco for $275,000.[28] The syndicate paid a portion of the purchase cost up-front, and owed the rest in installments of $25,000 at 12 per cent interest, to be paid every three months. The new owners were not interested in running the sawmill and so they leased the operation back to Heatley for some $20,000 per year.[29] Hastings Saw Mill continued to produce lumber, but the operation and its timberlands had acquired a new value in the real estate market.

By the summer of 1887, there were problems regarding the mill's ownership and the facility temporarily closed. There was a dispute between the British Columbia owners of the mill and Heatley, who considered the rental fee too steep. Negotiations ensued, and in the end the company was reorganized, with Heatley and his Dickson and DeWolf associates, of London and San Francisco, controlling half the company, and British Columbia, largely Victoria, investors owning the other half. The CPR also became involved in helping reach a suitable arrangement. The railway company promised to build a spur line into the site to allow the Hastings mill access to the rail trade. The agreement stipulated that the mill property had to be used for wood manufacturing, otherwise the real estate was to be divided, with two-thirds going to the owners and one-third to the CPR. The railway company did not want the sawmill firm to get involved in real estate development. A deal was struck; the mill began operating on 8 August 1887, and the rail link was completed in September. The 85,000-foot-capacity mill now had access to both the waterborne trade and the rail trade. In September 1889 ownership and management of production changed hands. The operation became part of the expanding empire of British Columbia entrepreneur John Hendry. Extensive changes were made to the facility, raising the mill's capacity to 150,000 feet per day.[30]

The other British merchant house prominently involved in the coastal lumber industry was linked to the sawmill at Moodyville, across Burrard Inlet from the Hastings mill. The Moodyville mill traced its origins to T.W. Graham & Company, who established Pioneer Mills on Burrard Inlet, five miles above the first narrows, in the summer of 1863. The mill shipped its first cargo of three-inch planks to New Westminster in August of that year.[31] Pressed by creditors, the owners decided to sell in December 1863, and the facility was purchased for $8,000 by Nova Scotian John Oscar Smith. Smith, a ship's carpenter, arrived in British Columbia in 1858, and had engaged in the grocery business and built and operated steamers.[32] Throughout 1864, Smith manufactured lumber for markets in Victoria, Nanaimo, and other local points. He even ventured into the export trade, shipping lumber and pickets to Australia in November 1864. The British Columbia Mill Company of New Westminster furnished a portion of the cargo.[33]

Smith, despite his foray into the export trade, was unable to keep the operation afloat; in January 1865, he sold to Sewell Prescott Moody, who, in association with Moses C. Ireland and James Van Bramer, paid $6,900.[34] The facilities were improved, and the now named Burrard Inlet Lumber Mills, owned by S.P. Moody & Co., began producing for its lumberyard in New Westminster, its dealer in Victoria, and the export trade. In March 1865 the company shipped 300,000 feet of lumber to Australia.[35] In 1865, the Moodyville mill loaded four vessels for the export trade; in 1866, five vessels; and, in 1867, seven vessels.[36]

In 1868 the company erected a new steam mill, some 300 yards to the west of the old one, which went into production in September. The new mill was capable of producing 50,000 feet of lumber per twelve-hour shift. There was also a new infusion of capital into the firm. Hugh Nelson and George Dietz, who had sold their interest in the British Columbia and Victoria Express Company in 1867, became partners. On 1 July 1870, the company changed its name to Moody, Dietz and Nelson.[37]

The merchant involvement in the Moodyville operation began with the activities of R.P. Rithet, a Victoria businessman, and Andrew Welch, a British agent located in San Francisco. In 1871 Rithet and Welch bought the business of J.R. Stewart, a Victoria businessman. John

Robertson Stewart was a merchant who participated in the lumber trade. From Edinburgh, he had arrived on the west coast in 1859, and by the summer of 1860 he was involved in business. Over the course of the 1860s, he developed connections with business interests throughout the world. His correspondents included Walker & Allen in Honolulu, S. De Beer in Melbourne, Graham Rowe & Co. in Lima and Valparaiso, and Ortiz Hermanos in Guaymas, Mexico.[38] Stewart imported goods from all over the world and exported lumber from British Columbia. By the late 1860s, Stewart controlled the Muir sawmill operation at Sooke. The Muirs were deeply in debt to Stewart. The Sooke mill was Stewart's 'most risky outstanding account' and, rather than call the debt, special arrangements were made between Stewart and the Muirs to keep the sawmill running.[39]

In 1871 Stewart was ill and eager to sell out, and the manager of the Victoria office, R.P. Rithet, was willing to buy. Robert Paterson Rithet was born in Scotland in 1844, and trained in a merchant's office in Liverpool before emigrating. He arrived in Victoria in 1862, establishing a wholesale grocery business. He also arranged lumber and spar consignments for the Moodyville sawmill, working with Gilbert Malcolm Sproat, whom he did not like, and Matthew Trottier Johnston, two former employees of Anderson & Co., now acting as agents on their own account. In the initial negotiations with Stewart, Rithet was associated with J.A. Raymur, another former employee of Anderson & Co. who became manager of the Hastings mill in the summer of 1870. However, Raymur felt the proposed deal was too costly and backed out.

Rithet then associated with Andrew Welch, a Scottish merchant in San Francisco; Rithet and Welch purchased the Stewart business in August 1871. Rithet was the British Columbia agent, and Andrew Welch the San Francisco agent; through Welch the company was linked to the trade of the Empire. The new owners also took on Stewart's worldwide connections, as well as the Muir account, which Rithet hoped to unload at a profit. Moreover, Welch had acted as the San Francisco agent for the Moodyville sawmill, and the new partners were keen to exploit this connection. Rithet wrote to Welch within weeks of the takeover that he had spoken to Hugh Nelson, a partner in the Moodyville company, and that Nelson was inclined to put their business in the hands of Rithet and Welch, adding that 'you might mention this to Moody perhaps in your letter to him. I am sure it can be arranged but we must be careful.'[40] Over the ensuing years the relationship between Rithet and Welch and the Moodyville mill became increasingly close.

In the 1870s there were ownership changes at Moodyville. Sewell Prescott Moody drowned on the night of 4 November 1875, when the ship he was travelling on from San Francisco sank. Hugh Nelson became the active owner and the mill manager. In 1878 the company was reorganized as the Moodyville Sawmill Company, and Andrew Welch became a major shareholder.[41] In 1882 Hugh Nelson sold his interest in Moodyville Sawmills to Andrew Welch, who became the sole owner of the firm. Moodyville was now completely integrated into Welch's commercial and resource empire. Welch was based in San Francisco, and owned a sugar refinery, as well as sugar plantations in the Sandwich Islands. His Liverpool connection was R.D. Welch & Co., and his British Columbia partner was R.P. Rithet.[42] The Moodyville sawmill was the largest in British Columbia, with a daily capacity of 85,000 feet.

The operation floundered in the mid-1880s. The mill relied on the local market, which could be serviced by scows, as well as the Pacific waterborne market. In 1887, twenty-three ships, destined for Australia (nine), China (six), South America (seven), and San Diego, took 16,566,008 feet of lumber from the mill, as well as 569,860 pickets, 14,091 laths, and 5 spars. The local coastal trade consumed 3,240,650 feet of lumber, 23,337 bundles of lath, and 23,926 lineal feet of poles. In response to an upswing, the mill underwent repairs and expansion in 1888 and 1889. However, the principal owner, Andrew Welch, died at the age of fifty, in July 1889, and the estate sold the company in 1891 to a group of English finance capitalists, including the Earl of Chesterfield, the Earl of Durham, Oliver Montague, and Arthur Heywood Lonsdale. The new ownership, however, could not compete with the depression of the 1890s, and in 1895 the provincial government seized the mill, which had considerable arrears in payments for timber.[43] John Hendry and his partners took over the operation in 1901.[44]

John Hendry, whose company absorbed both the Hastings and Moodyville companies, was the biggest success of the entrepreneurial age, and his achievement is striking for the longevity of his interest in coastal lumbering as well as the extent of his operations in the years before the First World War. Born in Gloucester County, New Brunswick, in 1843 or 1846, Hendry entered a milling family. His Scottish father, who had come to New Brunswick in 1832, operated flour and lumber mills. With the death of his father, Hendry took over the family business. He migrated west in 1872, working for sawmills in the Puget Sound region before becoming a superintendent overseeing the reconstruction of the Moodyville sawmill after a fire in 1874. Hendry then tried his luck in Winnipeg, returning to the west coast in 1875, building a small sawmill for W.J. Armstrong in New Westminster. In the fall of 1876, he formed a partnership with David McNair, from Restigouche, New Brunswick, and they erected a sash-and-door factory in Nanaimo, on Vancouver Island. Two new partners, Andrew Haslam and R.B. Kelly, both from Albert County, New Brunswick, were taken in, and in the same year they moved the machinery to New Westminster. The company was called Royal City Steam Planing Mills. A small sawmill with a capacity of only 7,000 feet per day cut lumber, which was used in sashes and doors, as well as boxes for the salmon-canning industry. In 1880 they sold this small mill to Gray, Hoy & Co., who established a local operation at Emory City. A new larger sawmill with a capacity of over 20,000 feet per day was constructed in New Westminster, and the company reincorporated as Royal City Planing Mills.[45]

In 1883 the New Westminster mill was enlarged to produce more than 30,000 feet per day, and a steamer was built to run between the mill and Port Moody.[46] The company also expanded to Nanaimo. They purchased the bankrupt Nanaimo & Wellington Sawmills, installed new machinery, and began operating in October 1883. Andrew Haslam took charge of the Nanaimo operation, which had a capacity of 14,000 feet per day.[47] In 1885 Royal City Planing Mills purchased the Dominion Sawmill Company. This company was established by New Westminster businessmen in 1883, hoping to take advantage of the boom associated with the construction of the Canadian Pacific Railway. The principal shareholders were John B. Henderson, James Gray, John A. Webster, all of New Westminster, and Alexander A. Green of Victoria.[48] The company took over Webster's mill, and new machinery increased the capacity of the mill to 45,000 feet per day.[49] They were soon shipping wood products to Yale, Lytton, Ashcroft, and Kamloops, and they had a lumberyard in Victoria.[50] The company, however, was unsuccessful. R.P. Rithet and Captain John Irving bought substantial shares in the firm on 1 September 1884, but the infusion of capital was insufficient to counter the general economic downturn. In the spring of 1885, the company was in the hands of the sheriff, and the property was sold in June. The operation was purchased by Royal City Planing Mills, whose New Westminster operation was adjacent to Dominion Sawmills.[51]

As we have seen, Hendry was instrumental in opening up New Westminster to the waterborne export trade in the late 1880s and in having his company purchase the Hastings mill in 1889. Anticipating the growth of Vancouver, Royal City Planing Mills established a sawmill with a capacity of 30,000 feet there on False Creek in 1886. The various enterprises of John Hendry, including the mill on False Creek, the Hastings mill, and the New Westminster operation, were merged into a new corporate entity, the British Columbia Mills Timber & Trading Company, on 28 March 1891. In 1905 the Hastings mill had the capacity to produce 200,000 feet per day, and the Moodyville operation, which was now part of the Hendry empire, had a capacity of approximately 100,000 feet.[52]

By 1909 John Hendry and his partners were preparing to sell off their vast holdings. The mill sites in Vancouver, New Westminster, and on Burrard Inlet derived their high value as pieces of real estate, not as factories producing lumber, and because of population growth and railway building this value was increasing. An assessment of the holdings in November 1911 made this clear: 'The Hastings Mill site had increased over $1 million in value within three years; Hendry's False Creek property had more than doubled; Moodyville had increased 40 per cent; the most spectacular increase was in New Westminster, where the plans of Mackenzie and Mann had caused Hendry's mill site to increase 600 per cent.'[53] In 1912, the False Creek mill site was sold to the British Columbia Electric Railway for $625,000, and the New Westminster mill site was sold to Mackenzie and Mann's Canadian Northern Railway for $700,000.[54]

Another nineteenth-century entrepreneurial enterprise was put together by the DeBeck family, but they sold out to outside capitalists in 1890. Like the partners in Royal City Planing Mills, the DeBecks were from New Brunswick. The DeBeck family began handlogging and farming in British Columbia about 1868. The family patriarch, Ward DeBeck, died in a logging accident in 1870, but the sons continued logging, taking out timber mainly for the Moodyville sawmill. By 1875 the DeBeck brothers operated a primitive sawing operation at New Westminster, and construction began on the more substantial 10,000-feet-per-day mill at the confluence of the Brunette and Fraser Rivers in 1877. The mill was running by March 1878, and included a wharf on the Fraser and a bridge over the Brunette. The mill was connected to New Westminster by a plank road. In March 1880, Clarence DeBeck sold his share in the firm to James B. Kennedy and, reflecting the change, the company was renamed DeBeck Bros. & Co.[55]

In March 1882, DeBeck Bros. upgraded their facility, and an observer noted that 'these mills are now ready for the railway "boom."'[56] In the fall of 1882, the mill was busy producing lumber for CPR bridges and other railway works, running two shifts. DeBeck Bros. established a selling agency in Port Moody, the terminus of the CPR, a growing community and headquarters for western construction. Further improvements in the winter of 1883–4 increased the capacity of the mill to more than 30,000 feet per twelve-hour shift.[57] The company became a joint-stock company in 1886, capitalized at $75,000, and changed its name to the Brunette Saw Mill Company. With a rail spur connecting the mill to the CPR and a lumberyard in Vancouver, the operation expanded. In 1889 a new sawmill with a capacity of 100,000 feet was built.

In 1890 the company was taken over by central Canadian investors. A. Barnet and Denis Murphy of Ottawa bought controlling interest in the company for $75,000, and in 1894 the capital stock of the company was increased to $300,000. J.M. McRae, president of the Ottawa Electric Railway Company, was one of the largest shareholders. Brunette Saw Mills also entered the export trade. By 1894 ships were loading lumber at the mill, and a new wharf, which allowed three vessels to load at one time, was built in 1895. The company, in association with San Francisco partners, purchased a barquentine, the *Wrestler*, in 1895 to ply the China trade.[58]

Andrew Haslam was another nineteenth-century lumberman, but his business ended in failure. In the mid-1870s Haslam joined Hendry as a partner in Royal City Planing Mills. After the company purchased a Nanaimo mill in 1883, Haslam left the company, taking full ownership of the Nanaimo operation as recompense for his interest in the company. He moved to the Vancouver Island city to run the operation, which increased its capacity from 14,000 feet in 1883 to 70,000 feet in 1890.[59] Andrew Haslam's Nanaimo Saw Mills was in financial trouble by the turn of the century. In 1901 he renegotiated his timber royalty payments with the provincial government, and in 1904 he was $26,600 in debt to the Royal Bank of Canada, which held some of his timber limits as collateral. By 1905 his overall liabilities were in the neighbourhood of $180,000; the company failed and was sold by creditors later in the year. The purchasers paid about $72,000 for the mill and timber holdings, and within a year turned over the properties for a profit of $65,000. Haslam took a job with the provincial government as the supervisor of log scalers.[60]

On Vancouver Island, William Parsons Sayward was a long-term participant in the lumber trade, even passing the business on to his son. In 1861, Sayward, originally from Maine, established a small lumbering facility at Mill Bay on the east coast of Vancouver Island, some 30 miles from Victoria. In the mid-1860s the Sayward mill was employing some twenty hands in the logging camp and mill.[61] Times were tough and W.P. Sayward kept afloat in 1868 by not paying the government its due timber rent for his Vancouver Island operation.[62] Sayward moved his milling operation to Victoria, and the mill itself was sold to Robert Dunsmuir in 1889. The private timber owned by Sayward was sold to local interests. Dunsmuir had sufficient timber of his own. Dunsmuir died in 1889, but his estate continued to own the mill until 1892, when it was purchased by a syndicate from Tacoma, Washington. In 1890 the Rock Bay facility, the largest in Victoria, had a capacity of 40,000 feet per day.[63] In the 1890s the mill operation was acquired by J.A. Sayward, the son of the mill's founder, and by 1907 the mill had a capacity of 150,000 feet. In 1909 it was sold to a U.S. syndicate based in Seattle. New partners from Michigan joined the firm, and the company became known as the Michigan Puget Sound Lumber Company. The company could not weather the economic downturn and went into liquidation on 18 May 1914.[64]

The economic boom from 1885 to 1892 brought a surge in entrepreneurs establishing milling operations and themselves in coastal British Columbia. Americans such as H.R. Morse, Hewitt & MacIntyre, and Theophilus Elford: Ontarians such as J.G. Scott, James Anderson, and William Lossee; and Maritimers such as William Munsie and George Cassady became involved in lumbering. However, a new note in the late 1880s was the arrival of two major lumber producers from outside the region who set up subsidiary operations in British Columbia. One was based in Canada and the other in the United States.

In 1887 James G. Ross and James Maclaren, prominent Ottawa Valley lumbermen, began purchasing timber rights in British Columbia, and in the spring of 1888 they bought a mill site, Senator Thomas R. McInnes's 100-acre farm on the north shore of the Fraser, just above Sapperton. The large sawmill, with a capacity of 200,000 feet per day, began operating on 19 May 1890, sending lumber out along the CPR. A year later, the Ross–Maclaren mills entered the waterborne export trade, sending a cargo of lumber to Australia on a ship that loaded at the company wharves. But production at the site was short-lived. By 1892 the facility was closed, a casualty of the economic downturn and the deaths of the two principal owners.[65] The operation remained idle throughout the rest of the 1890s. The firm also built a large sawmill at Barnet, a few miles west of Port Moody. Under the name North Pacific Lumber

Company, a mill with a capacity of 125,000 feet per day was built in 1890, and this plant continued to operate through the decade.[66]

The Sapperton operation, which had been idle since the early 1890s, was purchased by a U.S. syndicate headed by Lester W. David of Blaine, Washington, in 1904. Four years later the operation was sold to another group, which included members of the Swift family, the Chicago meat packers, as well Col. A.D. Davidson of Toronto and A.D. McRae of Winnipeg. In 1910 the Canadian Western Lumber Company, headed by William Mackenzie and Donald Mann, the Canadian railway magnates, and including Davidson and McRae, took over the operation.[67]

The U.S. business that expanded into British Columbia was the Victoria Lumber & Manufacturing Company, owned by the Humbird family of Wisconsin. They purchased the Croft & Angus mill at Chemainus in 1889. The Humbirds expanded the facility, and by 1890 the mill had a capacity of 50,000 feet per day. The onset of the depression, however, ended production, and the mill was idle for many years in the 1890s. After 1898, mill production was steadier; by 1906, the mill capacity had reached 320,000 feet. The Humbirds were also active on the west coast of the United States: the Humbird Lumber Company in Washington state involved Humbird, Weyerhaeuser, and Denkmann interests.[68]

The boom in the first decade of the twentieth century brought a new surge of investors in milling. The Canadian Pacific Lumber Company, owned by T.W. Paterson, Robert Abernethy, T.F. Paterson, and P.D. Roe, all from Ontario, purchased the Burrard Inlet Red Cedar Company at Port Moody in 1899, eventually selling out to a British corporation in 1910.[69]

Some local entrepreneurs, such as E.H. Heaps, built major operations. Born in England in 1851, the son of an architect and builder, he apprenticed seven years as a cotton broker in Liverpool, and then went to the United States for three years before returning to Manchester for eleven years. After operating his own business in Manchester, he migrated to British Columbia in 1889. In the 1890s he operated a Vancouver firm that sold supplies and equipment to the sawmilling and logging industries. The company represented central Canadian and U.S. manufacturers, claiming to have placed machinery in almost every mill and factory in the province, and at times traded in second-hand machinery. E.H. Heaps & Company was the selling agent for several mills and woodworking firms, with connections in Vancouver and on the prairies. In the 1890s, they also operated a small cedar mill which manufactured shingles and lumber. In 1902 the company purchased a mill site, and was soon running a sawmill at Cedar Cove, the largest mill in Vancouver, as well as a smaller mill at Ruskin, at the mouth of the Stave River. On 11 February 1905, the Ruskin mill burned to the ground, and on 6 October 1906 the same fate struck the Vancouver facility. The disasters forced retrenchment, and in 1910 E.H. Heaps, with the help of his four sons, operated a mill with only a 50,000-foot daily capacity. In July 1910, the Ruskin mill was again destroyed by fire, and Heaps went to London hoping to sell his Burrard Inlet holdings as real estate in order to raise capital to rebuild at Ruskin. In 1913 Heaps was at the head of a business empire that included trust, investment, fire insurance, coal mining, taxicab, and timber holding companies.[70]

In August 1910, a merger of four sawmill companies was put together, seemingly presaging a new era in British Columbia lumbering. The new corporation, the Canadian Pacific Lumber Company, was capitalized at $5 million and brought together four firms: the Canadian Pacific Lumber Company of Port Moody, the Anglo-American mill at Vancouver, the Gibbons Lumber Mills in the Kootenays, and a mill at Alberni. Five prominent Western Canadians sat on the board of directors. The company had big plans and expended much money in the next few years. However, the depression of 1913–14 caught up with the firm. It was soon bankrupt.[71]

While ownership patterns were complex, it is possible to make some observations on U.S. capital in the milling industry. Of the ten largest sawmills in the coastal region, only three were clearly owned by Americans in 1910. U.S. ownership tended either to be fleeting or to result in the U.S. owners settling in British Columbia. The Victoria Lumber & Manufacturing Company was an exception, as it was owned and controlled externally by the Humbird family from Wisconsin. Americans also tended to set up or purchase operations on Vancouver Island. This was likely because they could purchase timber outright in the privately owned Esquimalt & Nanaimo Railway Belt. Being unfamiliar with Canadian politics and policies regarding timber leases and licences, they wanted the security of freehold tenure of their timber.

American investment was also closely linked to rumours regarding U.S. tariff policy. When it seemed that Washington was about drop restrictions against the importation of Canadian timber, U.S. investors were much more interested in Canadian facilities. They wanted to export lumber made in Canada into the United States, a market that they understood and knew personally. The Humbirds invested in Chemainus at a time when lowering the tariff was a hot issue in U.S.

politics, and a Tacoma, Washington, syndicate bought a Victoria sawmill in 1892 because the members anticipated tariff revisions.[72]

To this point we have been discussing sawmills, but the shingle industry was equally dispersed, dominated by entrepreneurial and personal capital. Most of the big sawmill facilities operated shingle machines. In 1902 the British Columbia Mills Timber & Trading Company operations had a daily capacity of 160,000 shingles. E.H. Heaps & Company could produce 180,000 shingles, and Brunette Saw Mills 75,000. But there were companies that produced just shingles. In 1888, for example, G.F. Slater produced 100,000 shingles per day at his False Creek plant, and in 1890 W.L. Tait built a sizeable shingle mill at False Creek that turned out 52,000 units on a good day. W.L. Johnson, who had worked at Royal City Planing Mills for ten years, six of them as foreman, established a shingle mill on Gambier Island in Howe Sound in 1889. In 1893 and 1894 the company was shipping its product to Victoria and Ontario, among other places. Business, however, was not rosy in the mid-1890s and he was forced to return to Royal City Planing Mills as a superintendent in 1896.[73]

In 1902, Thomas Kirkpatrick, the Spicer Shingle Company, the Canada Shingle Company, A.R. Welch, the Fraser River Lumber Company, the British Columbia Shingle Company, Cook & Tait, and the Chilliwack Shingle Company ran operations that had the capacity to produce between 75,000 and 160,000 shingles each per day.[74] A major shingle producer was James A. McNair, owner of the Hastings Shingle Manufacturing Company. His company expanded production into the United States. In 1904 they operated two shingle mills, a sawmill and planing mill in Vancouver, and had four branch plants in the state of Washington to jump the tariff wall. The total daily capacity of the six shingle mills was 2,250,000.[75] The independent shingle producers were largely owner-operators, but, as in the sawmill sector, there were dramatic influxes into the industry in boom periods. And, like the millmen, no shingle producer or small group of producers was able to dominate, control, and rationalize the industry.

Neither millmen nor the shingle operators were able to keep out new entrants in good economic times. With few barriers to entry, firms entered the industry relatively easily in times of prosperity, leading to overproduction, falling prices, and economic hardship for the industry as a whole. Unlike coal and hard-rock mining, no firm or group of firms was capable of controlling the resource, timber and logs. This was a feature of the industry until later in the twentieth century. In 1913 the coastal forest was still too big. Nor were there technological restrictions that, because of cost, limited access. Unlike smelting and underground mining, milling lumber and shingles was a procedure that men with little capital could engage in. Large operations produced more lumber, but the economies of scale were not so great as to make smaller operations uneconomic. Thus small operators coexisted with large manufacturers. Nor did a few firms emerge to control marketing and distribution. If the industry had depended on overseas exports in a few markets, a large firm or conglomerate might have rationally controlled this point in the distribution system. However, the size and diversity of the local market were sufficient to sustain a great number of small and medium-sized operations.

The range of firms in the coastal lumber industry was extensive. In small communities in the Fraser Valley and on Vancouver Island, lumber mills with just a few employees catered to the needs of townspeople and the rural population. The visits of the provincial Inspector of Factories to coastal mills over the course of 1910 give a sense of the wide variety in size of the milling operations. He rated the mills according to the number of employees. Thirteen firms employed 201 or more people, ten employed 101 to 200, seventeen engaged 51 to 100, and there were fifty firms employing 50 or fewer workers. If we look specifically at New Westminster and the Fraser Valley, there were five large firms. The massive Fraser River Lumber Company employed 550, Royal City Planing Mills 280, Brunette Sawmills 242, Canadian Pacific Lumber Company 200, and the North Pacific Lumber Company 200. At the other end of the spectrum, twenty lumber-producing firms employed 50 workers or fewer. For example, the Patterson-Eckert Lumber Company of Chilliwack had 28 employees, the Jervis Inlet Lumber Company at Ladner 50, the Dominion Lumber Company at Aldergrove 25, the Mitchell Lumber Company at Mission 20, and the Silverdale Sawmill Company at Silverdale 40. Shingle mills operated with even fewer workers: the Essex Mill Company at Short Reed employed 8, the Port Kells Shingle Company at Port Kells 10, and the Maple Leaf Shingle Company at Clayton 15.[76]

Despite the growth of large companies, the coastal lumber industry remained decentralized in 1913. While there was a diversity of firm types, including many small personal firms, entrepreneurial capitalists spearheaded the development of the industry.[77] The business strategies used by British Columbia lumbermen and shingle entrepreneurs in this competitive economic environment are the subject of the next chapter.

3

Business Strategies

As the previous chapters have shown, coastal British Columbia lumbermen operated in an uncertain economic environment. In the years from 1858 to 1885, lumber markets were lethargic, and operators struggled to maintain their businesses. The years after 1885 brought expansion and more buoyant markets, but the boom-and-bust nature of the industry led to new problems. Lumbermen pursued different strategies to cope, and at times prosper, in these two economic climates. In the pre-1885 era, diversification was one strategy to maintain financial viability. Diversification meant, on the one hand, following a pattern established by the Hudson's Bay Company, where lumbering was only one aspect of broader economic activities. Companies engaged in non-lumbering activities to generate income in years when lumber markets were soft. Few companies in the colonial period felt that relying solely on lumbering was a wise course of action. On the other hand, diversification meant creating and selling a wider variety of wood products in order to appeal to a broader market. This was especially evident as a response to the growing local market in the 1870s and 1880s. The export mills had another strategy: beginning in the 1860s, they sought special financial treatment from governments in the form of tariffs and bonuses, arguing that the viability of their business demanded it. In the boom-and-bust economy after 1885, the main strategy of lumber and shingle operators was to fix prices and control outputs. Associations were formed to restrict and control competition, and cooperation in pricing and production levels was seen as the key to prosperity for all.

Diversification was evident in the coastal lumber industry from an early date. As Richard Mackie has shown, in the first half of the nineteenth century the Hudson's Bay Company moved beyond the fur trade to become a diversified, resource company that traded in salmon, lumber, cranberries, and dogfish oil, among other things.[1] The Hudson's Bay Company constructed a sawmill on Millstream, at the head of Esquimalt harbour, in 1848, and another at Nanaimo in 1854.[2] John Muir and his sons, former coal miners, operated a sawmill operation at Sooke by 1853, but they also engaged in farming and gold mining in the 1860s.[3]

At Alberni, Edward Stamp and Anderson & Co. operated a sawmill but still were very keen to develop agriculture, fishing, and mining. In August 1861, Stamp wrote that farming operations at Alberni were in full force and that they had made their first shipment of cured fish to Callao, while in February 1862 Stamp was requesting exclusive rights to mine copper on an island in Barclay Sound. Exports by the company in 1862 and 1863 included barrels of salt fish, fish oil, and skins and furs. Wood products were by far the most important product, but it is clear that the company saw diversification as an important business strategy.[4] In 1864 Anderson & Co. wanted to be appointed emigration agents for British Columbia in London and were staking claims for both coal and copper deposits.[5] After the sawmill closed, Anderson & Co. retained an interest in agricultural immigration, landholding, fishing, and mining in the Alberni area.[6]

Stamp, when he was involved with the Burrard Inlet mill, pursued interests, such as operating a postal service between New Westminster and San Francisco, and mining in the Chemainus District, that went beyond lumbering.[7] Mining was perceived as the best vehicle for attaining wealth in the colonial era, and Sewell P. Moody, George Dietz, and James Van Bramer, principals in the Moodyville sawmill, were involved in the British Columbia Coal Mining Co.[8] with other prominent mainland businessmen, such as W.J. Armstrong, John Pickavent Cranford, James Cooper, and J.A.R. Homer, who also had links to the lumber industry.

Diversification also came in the form of new product lines. In the 1860s colonial sawmills as a whole produced rough and dressed lumber, fence pickets, lath, tongue-and-groove flooring, and shingles. Even in the 1880s, the Burrard Inlet export mills still shipped mainly only rough and dressed lumber, pickets, and bundles of lath.[9] At New Westminster, however, enterprises catering to the local market introduced new products in the 1870s. The rise of the salmon industry was instrumental here. In August 1882, Royal City Planing Mills shipped 10,000 salmon cases to canneries on the northwest coast, and the construction of fish cases was an important aspect of the firm's business. Slabs rejected in the sawmill passed through a series of saws to create bottoms, tops, and sides, and there was a printing machine stamping the trademark of the particular cannery on the ends. One commentator noted that the making of fish boxes was very economic, as 'the proprietors are enabled to manufacture all the slabs hitherto wasted, which is a very important consideration as saw-logs become scarcer and dearer.'[10] The company also produced net floats for the fishing industry. DeBeck Bros., too, turned out boxes for the canneries.[11]

Mills expanded in other areas as well. Royal City Planing Mills, and other operations in Port Moody and New Westminster, installed machinery to produce shingles. Royal City Planing Mills was innovative in linking the production of lumber with a sash-and-door manufactory. Indeed, Royal City Planing Mills was originally a sash-and-door factory with a sawmill as an adjunct. In 1883 the manufactory was turning out 100 sashes and 100 doors per day.[12] The ill-fated Dominion Sawmill Company manufactured bedsteads, tables, cradles, and wooden vases, as well as sashes and doors, in its New Westminster facility.[13] Some larger sawmills also ran stores that sold goods to the community as well as to employees. The Hastings Saw Mill store sold logging contractors and ship captains equipment and tools, and supplies such as flour, tobacco, soap, blankets, clothing. Local settlers also patronized the mill's store.[14]

In the boom period from 1885 to 1892, many of the sawmills in New Westminster, Vancouver, and Victoria, which were close to urban markets, included sash-and-door manufactories as part of their operations. Royal City Planing Mills had a substantial manufactory in New Westminster, and Brunette Saw Mills added a sash-and-door facility in 1890. Royal City Planing Mills opened a sash-and-door plant at their False Creek sawmill in the spring of 1887, and Commercial Mills did the same in 1888. H.R. Morse's Michigan Lumber Company acquired a sash-and-door manufactory from Donahue and Boutelier in 1890, and, in 1894, J.B. Maclaren and H. Depencier, owners of the North Pacific Lumber Company at Barnet, established the Vancouver Sash and Door Company on False Creek. In Victoria, the Sayward sawmill was also associated with a sash-and-door facility.[15] In the rising lumber and shingle market after 1898, the close connection between milling and the manufacture of sashes and doors continued. Companies, such as Michigan Puget Sound Lumber, Genoa Bay Lumber Company, and the Hastings and Royal City Planing Mills operations of British Columbia

Mills Timber & Trading, all had extensive sash-and-door facilities in the years before the First World War.[16]

Beyond diversification, nineteenth-century lumbermen sought direct aid from the government to sustain their enterprises, beginning in the colonial period. The tactics of British Columbia lumbermen reflected the behaviour of businessmen across North America in this period. The provision of financial breaks by governments for particular businesses was a part of nineteenth-century Canadian business and politics. Businessmen lobbied for special treatment because of the ostensibly unique difficulties of their trade, and governments, keen to assure the viability of industries for economic development, responded with tariff breaks, as well as targeted bonuses and bounties. Town, colonial, provincial, and federal governments all allowed bonuses at times in the nineteenth century, and companies in Canada engaged in railway construction, shipbuilding, iron and steel manufacturing, textile production, and mining, among others, benefited from the schemes.[17]

In this tactic, British Columbia mill operators were surprisingly unsuccessful. In 1867 millmen clamoured for a protective tariff against U.S. lumber. John Robson, editor of *The British Columbian* and booster for the lumber industry, called for a duty of $2 per thousand feet on rough lumber and $4 per thousand feet on dressed lumber: 'As the matter stands at present the Colony presents an open market where the foreign manufacturer can dispose of his lumber without contributing a cent towards our revenue.'[18] The proposal was deemed fair as it was similar to duties imposed on colonial lumbermen by the U.S. government. The legislative council recommended a new tariff schedule on 14 February 1867, whereby $3 per thousand (M) would be charged against rough fir or cedar and $5 per M on dressed fir or cedar, as well as $1 on laths per M and $2 per M on fence pickets.[19] However, the new tariff schedule also offered protection to colonial farmers who raised crops and livestock used in lumbering, and raised the cost of goods from the United States necessary to the mills. Thus lumbermen's costs increased. The tariff debate exposed the conflict between lumbermen and farmers in the colony. Farmers complained that the mills imported their hay and other agricultural supplies from Puget Sound. The millmen countered that the supply of hay and produce in the colony was insufficient. Moreover, while the 1867 protective duty of $4 per ton of hay had been imposed with the intent of stimulating production, by the spring of 1868 lumbermen were complaining that there was still a shortage of hay in the colony and the mills had to pay more for hay from Puget Sound.[20]

The lumber interests first responded to the new tariff by requesting exemption from the schedule for operations that competed with Puget Sound sawmills. They argued that lack of access to the U.S. market and lower costs for food, labour, and provisions for Puget Sound competitors justified the exemption. A resolution was introduced in the legislative council on 6 March 1867, recommending that foodstuffs such as sugar, salt beef, and bacon, as well as such things as iron, brass, copper, and India rubber belts, which were necessary for mill machinery and repairs, be admitted into the colony duty-free.[21] After discussion it was decided that this measure was inexpedient.

Thereafter the millmen began to plump for a bonus or drawback on lumber exported from the colony. On 22 March 1867 the council resolved to recommend to the governor that a sum not exceeding $0.50 per thousand board feet on all sawn lumber exported from the colony be granted to the mills.[22] However, there was no reaction from Governor Frederick Seymour. In January 1868, Edward Stamp wrote to the acting colonial secretary, requesting that the drawback scheme be implemented. He stated that, if the measure was not granted, 'every saw mill in the country will have to discontinue working.'[23] The government was not sympathetic and the governor vetoed the proposal. The government's position was clear:

> I view artificially cheapening the products of a country as a palpable mistake. I am as much opposed as ever to the principle ... The mill company must be in a bad way indeed, if their business can only be carried on by the rest of the population being taxed for their support. Besides if a bounty be paid on lumber exported, what arguments can be *advanced* against a similar payment in respect of every ounce of gold *produced,* or every ton of coal, or barrel of salmon exported.[24]

Stamp's mill was in a bad way, and even later pleas from the chairman of the British Columbia and Vancouver Island Spar, Lumber, and Sawmill Company, C.W. Fitzwilliam, could not move the government.[25]

In later years the export mills were more successful in persuading the provincial government. Section 33 of the 1888 provincial Land Act provided for a rebate on royalties on exported lumber: half of the timber royalties owed the provincial government would be returned to the companies on timber products that were manufactured in British Columbia mills and sent beyond provincial borders. An order-in-council rescinded the policy on 1 June 1891, but the depression caused the government to reconsider. In 1893 lumber prices in export markets dropped from $11 to $7 per thousand, and on 1 March 1893 export rebates were reinstated. The new policy provided for a rebate of $0.25 per thousand board feet on lumber shipped to foreign ports, retroactive to 31 August 1891. In 1893 nine companies qualified for the rebate, including Moodyville Sawmills ($11,625.80), Hastings ($6,512.31), Victoria Lumber & Manufacturing ($1,031.75), and H.R. Morse ($499.41).[26] In the 1890s the federal government also offered a rebate on lumber produced from Dominion timber that was exported to places other than Manitoba and the Northwest Territories. In 1896 this amounted to a rebate of $0.40 per thousand feet against dues owed the federal government.[27]

By the 1880s, lumbermen were increasingly beginning to look to their own devices, creating associations to cope with the interconnected fluctuations of markets, production levels, and prices. At a secondary level, associations were vehicles to fight for lower freight rates, lobby against the unsatisfactory tariff policy of the federal government, and coordinate anti-labour measures.

The lumbermen's flight from competition was hardly unique to the West Coast. Despite the much-vaunted celebration of nineteenth-century businessmen as competitive, independent actors in a free market, the historical reality was quite different. Businessmen viewed uncontrolled competition as an evil, a relentless process that drove down profits and undermined the entrepreneurs' chance of sustaining a decent standard of living. As workers increasingly organized into unions in the last half of the nineteenth century to further their interests by acting collectively, business people, too, formed associations or combines to represent their interests. In the 1880s and 1890s these associations were ubiquitous. At an 1894 hardware convention in Toronto, for example, the Wire Nail Association, the Wire Association, the Screw Association, the Bolt and Nut Association, the Rivet Association, the Barb Wire Association, the Bar Iron Association, the Cut Nail Association, the Horseshoe Association, the Tack Association, and the Paint Grinders' Association were all represented.[28]

A major goal of associations was to set or fix prices that would allow members what they considered a decent profit margin. Associations also lobbied governments, attempted to set production quotas to maintain price levels, and sought standardized wage schedules to defeat attempts by workers to raise wages and thus keep expenses down. Despite the initiatives, success was difficult to achieve. There were always one or two members of the trade who refused to participate, forcing the others to respond in order to retain market share. Moreover, public opinion, the voice of consumers, signalled outrage at the collusion between businesses. This forced political action, and in 1888 the federal government struck a committee to investigate combinations in restraint of trade for the first time.

Business cooperation was not only a central Canadian phenomenon. There is evidence of lumber operators in the U.S. Pacific Northwest trying to work together as early as the 1850s, but it was not until the 1890s that these attempts took on a degree of seriousness.[29] According to John H. Cox, price-fixing, setting production quotas, and freight rates were at the heart of these associations, while other goals such as uniform grading, employees' wages, and tariff lobbying were secondary. Cox concludes that 'concerted action through the associations worked well in good times, but collapsed when "hard times" struck the industry.'[30] William G. Robbins is more critical: 'During the great heyday of lumber-industry expansion, from the 1880s to 1907, efforts to cooperate through associations, pools, and dividing markets failed miserably.'[31] British Columbia lumber and shingle producers fit nicely into this continental pattern.

Cooperation was necessary not only in weak market conditions. When the lumber and shingle economies were hot, rising demand and prices stimulated the construction of new saw and shingle facilities. This expansion of the number of suppliers, coupled with the increased production of existing plants, led to fierce competition, overproduction, and unbearable price levels. Firms went broke in boom periods. Commercial Mills on False Creek went bankrupt in 1889, citing 'the lowness of prices and overproduction for local supply.'[32] Edward H. Heaps summarized the uneasiness of millmen in the expansionary year 1909: 'Speaking generally, I figure out that the unsatisfactory condition, and there is one, is due to overproduction mainly, to want of margins, to our long distance from the market, to our high freight rates and other differences, such as competition from a supply of lumber much nearer to what is our only large and our natural market.'[33]

It should also be noted that, in periods of ostensible prosperity, uncertainty still ruled. For example, although the period after 1898 was one of general prosperity for lumber operators, from the perspective of the producers at the time these markets were never stable. There were yearly, and even seasonal, fluctuations. In general, slow market expansion from 1898 to 1901 turned into a boom from 1901 into 1904. This was followed by a downturn through 1904 and 1905, but another substantial upturn occurred in 1906 and 1907. The San Francisco earthquake and fire in 1906 created a great demand for lumber, and the temporary suspension of the duty against Canadian lumber allowed coastal British Columbia mills to contribute to the rebuilding of the California city. The market slackened in late 1907, and by early 1908 mills were closing as a result of a lumber glut. The fortunes of the British Columbia Mills Timber & Trading Company reflected the general trend, and general manager John Hendry passed the bad news to his partners in his report at the end of 1908: 'The Balance Sheets and Accounts for 1908 ... do not, I regret to say, show as satisfactory a result as that for 1907.'[34] Recovery in 1909 ushered in a period of rapid market expansion, and the industry achieved record production levels until the summer of 1913, when the prairie market collapsed. The Thurston–Flavelle sawmill at Port Moody had posted an average of 28 per cent profit on sales between 1909 and 1912, but in 1913 this dropped to 12.2 per cent, and in 1914 to 6.3 per cent. One of the owners, Aird Flavelle, remembered: 'It nearly broke us, and then the bank called our loan.'[35] In the summer of 1914, British Columbia mills were cutting only about one-third of the logs they had consumed in previous years, and only 50 of 240 logging camps operating in 1913 were still open.[36] It was not until 1915 that the coastal lumber economy rebounded. Markets, in good and bad times, were a concern for businessmen.

The beginnings of organized cooperation in the coastal lumber industry can be traced to the 1880s. On 6 March 1886 sawmill operators from New Westminster and Port Moody met in the Royal City Planing Mill offices in New Westminster, and as well as discussing matters of general interest to the millmen, they adopted a schedule that raised the price of lumber. They also appointed a committee to 'draft by-laws and do other work preliminary to forming a permanent organization.'[37] While an organization was not established, a year later representatives from Hastings, Moodyville, and Royal City Planing Mills met and decided again to act in concert to raise the price of their products.[38] In late 1887 the export mills on Burrard Inlet also participated with U.S. mills in an association designed to establish a price schedule for all lumber shipped from North America to Pacific markets.[39]

The depression of the 1890s made the situation more pressing for mill owners. In Victoria, local lumber dealers came together in the spring of 1892 under the name Victoria Lumber Exchange. They set a price schedule, agreeing to sell rough lumber at $10.00 per thousand board feet, best-quality dressed lumber at $18.00 per thousand, second-quality dressed lumber at $14.00, and laths and shingles at $2.25 per thousand. However, within weeks one firm broke ranks, refusing to adhere to the scale, and the organization was in shambles. In a few months, though, the association revived, and the price of rough lumber was advanced to $12.00 per thousand. But by October there was again trouble: 'the combination was a farce,' with all dealers cutting prices. An American-owned firm refused to join the Lumber Exchange, and by setting its own prices destroyed the association. By February 1893, due to lack of organization and the deepening of the general depression, the price of rough lumber in Victoria dropped to $7.00 per thousand board feet.

In March another attempt was made to organize, and the price of rough lumber was advanced to $10.00 per thousand, but by early August this initiative collapsed, and the price returned to $7.00. Victoria mills, however, persisted. In the spring of 1894 a new scheme was hatched, and this time building contractors were included in the organization. The price of rough lumber was advanced to $8.00. The fate of this organization is unknown, but it is fairly safe to say that it was short-lived.[40]

Mainland organizations suffered a similar fate in the 1890s. Seven mills came together in May 1892 to establish a rate schedule. However, the Brunette Saw Mill Company soon became dissatisfied, claiming that some of the other participants were not adhering to the rates set by the combine. In July, Brunette Saw Mills gave the requisite three months' notice that they were leaving the organization, and in October the organization collapsed, the mills reverting to setting their own price schedules. In the summer of 1895, during a revival of the lumber economy, the millmen decided it was time to advance prices. A conference held in Vancouver on 3 July 1895 attracted representatives from most of the mills in the province. Since the selling price of lumber had been little above cost in the previous three years, with most mills running at a loss, and since the mills on Puget Sound had just advanced their prices, the meeting decided to increase prices from $2 to $3 per thousand board feet in anticipation of a more buoyant market.[41]

The shingle industry established a province-wide organization in late 1893. In discussions throughout 1893, shingle manufacturers argued that a combination was needed 'to keep up prices, which ... have been cut below cost by too active competition.' The British Columbia association advanced shingle prices, but was dead by August 1894. In 1897 talks began again, and the British Columbia Shingle Manufacturers' Association was formed in October. Their November meeting was called specifically to fix prices.[42] The difficulties of price-fixing were obvious. In order to be successful, all market participants had to adhere to the established schedule, and just one maverick company would undermine the whole agreement. As individual operations had their own particular financial deadlines or difficulties, and as some owners saw opportunities to increase their share of the market by dropping prices, these attempts at controlling prices had no lasting impact.

The dropping of the U.S. tariff against Canadian lumber in the 1890s drew Canadian producers into U.S. combines. As the Canadians now had access to the United States, in competition with U.S. producers, they had to be persuaded to adhere to regulations that governed the industry. The Americans had approached British Columbia millmen in 1893 in an unsuccessful attempt to form a lumber combine on the Pacific coast. The new tariff and depressed lumber market in 1894 and 1895 made the inclusion of the Canadians imperative. Hastings Saw Mill was singled out as particularly obnoxious for undercutting prices in the San Francisco market: 'The Hastings Mill on Burrard Inlet made a number of sales, some as low as $7.25 a thousand. Eight vessels were soon loading at its docks and the mill was preparing to run night and day. The mill was so busy that it had to give up at least one cargo to the neighboring Moodyville mill ... In his attempts to keep his firm solvent, the owner of the Hastings Mill helped pull down prices for all.'[43]

The desperate circumstances of 1895 led to the creation of the Central Lumber Company, which included sixty cargo mills in California, Oregon, Washington, and British Columbia. There were five participants from coastal British Columbia, including the Hastings operation. The goal of the organization was to stop 'the cutting and slashing of prices, which has been going on for several years, rendering the business profitless.'[44] Victoria Lumber & Manufacturing refused to join the combine, reopened its Chemainus facility, which had been closed for three years, and began undercutting the prices set by the Central Lumber Company with much success. Other problems plagued the organization and it collapsed in late 1896.[45]

Regulating prices and output was a major goal of the British Columbia Lumber and Shingle Manufacturers' Association, an organization that brought together both lumber and shingle operators in 1898. A lumber trade journal commented in 1900 that the association aimed to regulate prices 'so as to enable the mills to carry on business much more profitably than in the past, and to more equitably divide the business ... Owing to the friendly relations now established, ... the output will be regulated to some extent so that all the mills in the province will as far as practicable be made to share alike, in all the business offered.'[46] By the December 1901 annual meeting, the majority of the coastal mainland mills had joined, and committees were struck to deal with a range of issues. These committees were the lumber mill committee, the shingle mill committee, the sash-and-door committee, the railway rates committee, the labour committee, and the log committee.[47] There were interests beyond price-fixing and setting quotas, but, except for tariffs, they were secondary.

Still, coordinated action was not an easy feat for either shingle or lumber producers. A series of attempts began with much fanfare, but the collapse of agreements was often ignored in the media. In 1901 a group of shingle manufacturers agreed to follow production and price guidelines. The Pacific Coast Lumber Company agreed to shut down its shingle mill and act as the selling agent for four other shingle producers. The arrangement, however, was not sustained. In 1903 shingle overproduction created a glut in Manitoba and the Northwest, and the British Columbia Lumber and Shingle Manufacturers' Association members agreed to close their shingle mills on 15 August for a period of thirty days.

In January 1904 the major lumber producers in the Lower Mainland, including British Columbia Mills Timber & Trading, Pacific Coast Lumber Company, E.H. Heaps & Company, Brunette Saw Mill Company, and Canadian Pacific Lumber Company, agreed to follow a price list established by the British Columbia Lumber and Shingle Manufacturers' Association. The memorandum they signed began as follows: 'We, the undersigned, hereby covenant, one with the other, and solemnly promise that we will maintain Local Price List Four (4) and the Association Terms of Sale, – and not evade its letter, spirit or terms in any manner, directly or indirectly.'[48] A month later, as markets weakened, members of the association entered a pact to manage production and allocate the local lumber trade. They set up a joint stock company to market the production of the member companies. British Columbia Mills Timber & Trading was given 30 per cent of the market, E.H. Heaps 17 per cent, Pacific Coast Lumber 15 per cent, Robertson & Hackett 14.5 per cent, Rat Portage Lumber 13.5 per cent, and Urquhart Mills 10 per cent. However, the pool project was abandoned because one large firm, E.H. Heaps & Company, refused to participate in good faith. The other mills could not hold up prices, and by September 1904 lumber was being sold at prices up to 35 per cent below those set by the association.

L.A. Lewis, manager of the Brunette Saw Mill Company, confessed that there was also an attempt in 1905 to set prices: 'I admit frankly that we did everything our brains could think of to establish a uniform price among the millmen, that we would sell at one price, so as to put the lumber industry on a stable footing.'[49] In the spring of 1906, another attempt at price-fixing was initiated when coastal sawmill companies set a new rate schedule, raising the price of lumber from between $1 and $5 per thousand board feet, depending on the grade of the product.[50]

Consumers in British Columbia were suspicious of the activities of the lumber companies. In 1909, Thomas Turnbull, W. Dodd, and R.A. Stoney, representing the New Westminster Trades and Labour Council, voiced a popular perception in the province: 'There is at present a lumber trust in this Province which controls the price of lumber to the consumers to a very great extent and we believe that if the government grants the wishes of these gentlemen in a few years the result will be that the complete control of the lumbering industry will be centred in one corporation, and, therefore, competition will be completely eliminated.' Turnbull supported the position by referring to the price list produced by the British Columbia Lumber and Shingle Manufacturers' Association in January 1908. He added that, having purchased lumber in Vancouver and New Westminster for the past twenty years, he was well aware that the association price was followed.

The industry was sensitive to public accusations of collusion and price-fixing. Their legal, public position was that while there was indeed a price list distributed to members, it was merely a guide, and the mills were not bound to sell at the recommended price. However, Turnbull also noted that 'some things you could get cheaper but it had to be kept very quiet.' In other words, he acknowledged that mills often broke ranks.[51] Privately, many millmen wished that Turnbull was right, that the British Columbia Lumber and Shingle Manufacturers' Association was able to force all members to adhere to an established price list to bring a degree of long-term stability to the industry. The industry was engaged in price-fixing, but it was not as successful as the public imagined.

Prairie consumers were no happier with the lumbermen than were their Lower Mainland counterparts. In the prairies, coastal millmen battled farmers, as well as prairie retailers, the railways, and the federal government in their struggle to sustain the market. Public accusations of price-fixing, gouging, and undermining the development of the prairie West forced a federal government inquiry into the lumber business on the prairies in 1903. Justice Richards of Manitoba was appointed commissioner, but he returned his commission on 26 February 1904, before any evidence had been tendered.[52] In 1904 the British Columbia Lumber and Shingle Manufacturers' Association won a battle, when they charged the Canadian Pacific Railway with setting freight rates that discriminated against cedar lumber. The Railway Commission agreed with the lumbermen, issuing an order that railway companies should desist from charging a higher rate on cedar than was charged on fir, spruce, pine, and other lumber.[53]

Relations between coastal lumber producers and prairie retailers were especially heated. One grievance went back to 1902. In February of that year, a delegation of British Columbia millmen met with retail dealers in Winnipeg. A committee was appointed to interview Mr Whyte, vice-president of the Canadian Pacific Railway, to negotiate lower freight rates. An arrangement was made whereby the British Columbia producers agreed to reduce their price on lumber by $1.00 per thousand board feet and the CPR agreed to lower its freight rate by two cents, making the rate from the coast $0.38 instead of $0.40 per hundredweight. The CPR and the lumber producers made sacrifices, but the prairie retailers did not. The cost of lower lumber prices for consumers in the prairies was borne by the millmen and the railway.

During the slowing market and falling prices in 1904, lumbermen continued to bear the brunt of the hardship, but, according to the millmen, the retailers were unwilling to do their part by lowering their profit margins. When prairie farmers and consumers complained about the cost of lumber, millmen felt that they were unfairly blamed, because prices were being propped up by the retailers. The British Columbia coastal mills were also angry because the prairie retail association was buying cheaper U.S. lumber and selling it at high prices: 'in face of our repeated protests, [they continue] to purchase lumber from the United States, and although purchasing at a lower price, charge their customers the same prices as rule for lumber manufactured in Canada, thus working injury both to the settlers of the Northwest and the Lumber Manufacturers of British Columbia and Canada.'[54]

In 1904 lumber producers attempted to circumvent the prairie retail agencies with the establishment of the Manufacturers' Lumber Company, capitalized at $1 million, to sell lumber directly in the prairie market. The new marketing company included lumber companies from Ontario, Manitoba, Saskatchewan, and British Columbia. The largest number of participating mills were in the Kootenays, but the coastal region was represented by the Canadian Pacific Lumber Company, Brunette Saw Mills, the Ladysmith Lumber Company, the Shawnigan Lake Lumber Company, and the Victoria Lumber & Manufacturing Company.[55] On 1 January 1905, the coastal mills formally withdrew from the Western Retail Lumber Dealers' Association, and now no British Columbia mills had any connection with the organization. While the producers threatened to establish their own depot yards in the prairies, the proposition was easier to announce than fulfil.[56]

Prairie consumers continued to accuse the British Columbia lumbermen of gouging. The rapid rise in lumber prices in 1906 spawned charges of a 'lumber trust' in the United States, leading to a number of state and national inquiries and calls for 'trust-busting,' and in Canada the federal government launched an inquiry into lumber prices in the prairies.[57] The committee concluded in April 1907 that the retail dealers in the prairies were indeed engaged in a combination to unduly enhance lumber prices, but that although British Columbia millmen participated in combinations, there was insufficient evidence to show that their prices were excessive.[58] Coastal British Columbia manufacturers continued to try to circumvent prairie retailers and wholesalers. In June 1910 representatives of most of the Vancouver mills, as well as Fraser Mills and Victoria Lumber & Manufacturing, decided to put 50 lumberyards into operation in Manitoba and Saskatchewan by the end of the year. The plan was then to increase this number to 500 in the future. The reasons were familiar: the millmen felt that the prairie retailers had unduly advanced the price of lumber to consumers, thus leading to less consumption and an unnecessary drop in demand. The announcement prompted a meeting between British Columbia producers and prairie retailers in Banff, where a new price schedule was negotiated.[59]

In the early twentieth century, management of the prairie trade included the lumber producers of the Kootenay region of southeastern British Columbia. The Kootenay lumber industry began when the Canadian Pacific Railway was built through eastern British Columbia in the 1880s, but the boom in the prairies after 1898 and the completion of the Crow's Nest Pass Railway furthered development. By 1907 there were some sixty-four sawmills in the Kootenays, with a production capacity of 3.2 million board feet per day. In the coastal region there were fifty sawmills, with a daily capacity of 2.7 million board feet, that catered to the prairie trade. The two regions sent approximately the same amount of lumber to the prairie West.[60] While coastal mills produced higher-quality lumber and were able to operate almost year-round due to a favourable climate, the Kootenay mills had the advantage of cheaper freight rates because of their proximity to the prairies. The Kootenay mills had their own trade organization, the Mountain Lumber Manufacturers' Association.

There was tension between coastal millmen and their counterparts in the Kootenays. In the fall of 1907, coastal shingle mills agreed to curtail production because shingle stocks were too high. All coast mills decided to shut down through November and December to get rid of the surplus. Kootenay shingle producers, however, refused to cooperate, thus undermining the activities of the coastal operators. On 27 June 1908, coastal and Kootenay lumber producers met in Calgary and established a price agreement for the prairie market. The agreement was to last not less than one month, and to the dismay of coastal operators the Kootenay mills broke away on 1 August, just over a month after the agreement had been set. Coastal producers were forced to introduce panic selling prices to maintain their market share. Later in the year, the mountain millmen wanted the assistance of the coastal mills in order to forge a united front to raise prices. The coastal firms refused. They were bitter over the actions of the Kootenay producers in the past, blaming them for the prevailing low prices. Now in late 1908 the coastal millmen were 'disposed to give the mountain men an overdose of their own medicine.' In assessing the situation, an industry analyst mused on the gospel of cooperation:

> According to the latest reports the condition of affairs in the shingle business now is one of 'every man for himself.' It is a matter of regret that this should have come to pass, but in the light of events and of human nature it is difficult to understand how it could have been otherwise. The war has been going along in a leisurely fashion for some time, but now it seems to be 'to the knife.' If it must be so, the fiercer it is the better, so that it may be the sooner over. Industrial struggles between those whose interests should be identical is a great folly. The fact that the crisis has now been reached gives hope that in the natural course of events a healthier condition of affairs will soon be brought about.[61]

In the years from 1858 to 1913, coastal producers were not content to rely on lumber-market forces to decide the fate of their enterprises. In the weak, slow-growing markets from 1858 to 1885, they diversified their activities and generated a wider variety of products to sustain their companies. Although unsuccessful, export mills sought tariff and financial incentives from the colonial government. After 1885, like other Canadian and North American businessmen, they sought to manage prices and production in what they perceived as a hostile environment. Lumber and shingle production increased dramatically in coastal British Columbia in the first decade of the twentieth century, and much money was made. But for participants in the industry the atmosphere was chaotic. Booming markets brought new capital, more milling operations, and an increase in productive capacity. The federal government, railways, retailing organizations, and consumers pursued interests that conflicted with the goals of lumber producers, and in the end neither tariff relief nor market stability was achieved. There was no consistent, orderly cooperation between mountain mill owners and coastal operators, between retailers and producers, or even among coastal millmen themselves. During economic downturns especially, the time when cooperation was most essential, firms looked to their own interests, and any sense of the good of the whole was secondary.

4

Government Policies

Government participation in the management of the forests and the forest industry began early. From colonial days, government was intimately involved in the forest economy. Committed to the creation of a prosperous forest industry built by private enterprise, governments sought to implement policies that would further this goal. Most obviously, governments developed timber-allocation procedures. The colonial and later the provincial state had sovereignty over public lands and the forests, and they conceived the terms under which the rights to control and cut the forests were transferred to private interests. Governments in most instances did not alienate the forests, which would have allowed timberland to be owned outright in a freehold land-tenure system, but rather retained rights to the forest land. Thus the government retained an ongoing interest in forest policy, an arrangement supported by businessmen, politicians, and the public.

Most directly, timber leases and licences defined private access to cutting the forests, but the state also had a broader function. Licencing and leasing procedures were embedded in the larger structures of colonialism and capitalism. Committed to industrial capitalist expansion, the state recognized the need to create, sustain, and enforce the private-property rights of firms and individuals. Forest law had to be imposed and followed. In imposing the authority of the colonial and provincial state over forest land claimed by First Nations peoples, in surveying and legalizing private control of mill sites, in adjudicating disputes over timber among different loggers, in establishing rules that defined fairness in measuring the volume of logs cut which would satisfy both mill owners and logging firms, and in distinguishing the land rights of farmers and lumbermen, the state created a climate conducive to the capitalist exploitation of the forests.

The colonial Vancouver Island and British Columbia states, and the later provincial British Columbia state, were most crucial in developing the legal framework for commercial exploitation of the forests. Under the Canadian constitution, the provinces had jurisdiction over lands and forests. The federal government did gain control of some forests in the province, as we shall see, but the federal role was secondary, following patterns set by the province. Moreover, the federal government was rather lackadaisical in managing its interests in provincial forests. A focus on the colonial and then the provincial governments captures the main thrust of state management policies in this period.

To pave the way for commercial lumbering, the colonial government asserted the sovereignty of the British Crown in the region.[1] The establishment of the mill on the Alberni Canal was a messy business. This was Nuu-Chah-Nulth territory, and one group, the Tseshaht, had to be appeased to make way for the enterprise.[2] In March 1860, W.E. Banfield, government worker and associate of Captain Stamp's, estimated that there were some 550 First Nations people at Banfield, plus those members of the two other tribes that inhabited small villages on Barclay Sound. Banfield was suspicious of the Native people: in October 1859 he had recommended that a warship be sent to the west coast to strike fear into the hearts of the local people, and in January 1860 he had taken a Barclay Sound chief to Victoria to meet Governor Douglas. Douglas gave the chief a suit of clothes.[3]

On 29 June 1860, the first contingent of Stamp's workers arrived on the Sound, and Banfield, who now worked for Stamp as well as for the government, introduced the newcomers to the local people. Within weeks there was tension between Stamp's workers and the local people. The whites were very disheartened and frightened. After the arrival of another group from England on two armed vessels, the Europeans were forced to negotiate with the Tseshaht on 2 September 1860. Stamp gave the Native people molasses, food, and trinkets, and in his mind this seemed to serve as a treaty with them.[4] No formal government treaty was signed with the First Nations peoples on Barclay Sound.[5]

Peace was not achieved. In October the white settlers considered leaving the area, and Stamp wanted a vessel of war sent to cow the local people; he feared exposing his property, 'valued at little short of £10,000, to the depredation of the Indians.'[6] Sproat later noted that 'Governor Douglas was good enough to send H.M. gunboat *Grappler* to our aid, and the arrival of this vessel convinced the natives that the King-George men were in earnest.'[7] Stamp had himself appointed Justice of the Peace and began sending Native peoples to Victoria to serve four months' hard labour for theft: 'Thefts have become so common here with the Indians; this step was absolutely necessary; I was told by the chief that I had stolen their land and they had a right to steal from us.'[8] In 1865, Anderson & Co. was making claims to secure their landholdings in the Alberni area, and their representative, Gilbert Malcolm Sproat, noted that the company would not expect remuneration from the government 'for the value of a necessary second purchase of the land from the Indians, nor for the presents given to the Indians of Barclay Sound on various occasions, by which peace was secured in that quarter without trouble or expense to the colony.'[9]

On the mainland, too, milling and logging operations intruded into First Nations territories. On 1 June 1865, James B. Launders was engaged surveying the mill site for Stamp's operation on Burrard Inlet. He came to the conclusion that part of the 100 acres was the site of a Native village: 'the N.W. corner occurs in the centre of an Indian Village to clear which would only give the saw mill claim about 90 acres; by the appearance of the soil and debris this camping ground is one of the oldest in the Inlet. The resident Indians seem very distrustful of my purpose and suspicious of encroachment on their premises.'[10] Launders's superior interpreted the situation differently: he stated that a Squamish Indian named 'Supple Jack' had merely been squatting on the land for the past three years and that there were only two male relatives living nearby. Captain Stamp had no objection to their remaining where they were: 'They can be at any time removed; the ground does not belong to their tribe.'[11] The transfer of the land to Stamp proceeded.

The colonial government also had to devise a policy to turn timber over to lumbering companies. The proposal for the Stamp mill in 1859 made the making of a firm policy for large operations crucial. Companies needed access to sufficient timber to justify their investment in milling and logging operations. The government, for its part, wanted to encourage industrial activity, but at the same time it feared timber falling into the hands of speculators. The government also was leery of lumbermen controlling land that was suitable for agricultural settlement, since farmers were perceived as the backbone of a stable, flourishing society. In the British monarchical tradition the public lands were the property of the Crown. Military requirements in the age of wood and sail necessitated close control of the forests to meet the demand for masts, spars, and planking.[12] In the colonies of Vancouver Island and British Columbia, the Crown was recognized as the owner of the land and resources, by non–Native people at least, and through local governments dispensed land, timber, and other resources to private individuals for commercial exploitation.

Until Edward Stamp presented his demands for timber to feed his proposed Alberni mill in 1859, the colonial government had simple timber-allocation procedures. Early, small sawmill operations logged their own private land, neighbouring pre-emptions, and, on occasion, public land, and this timber supply was sufficient to meet their limited needs. In 1853 the colonial government passed a series of timber regulations. A duty of ten pence per load of fifty cubic feet was imposed on all timber cut on the public lands, and provided 'that no person not being a subject of her Majesty the Queen and a resident of Vancouver's Island, shall cut timber on the public lands under a penalty not exceeding £20.'[13] Governor Douglas explained that the regulations were 'altogether protective, it being thereby intended to prevent the waste and destruction of timber on the public lands, and to throw the timber trade, as much as possible, into the hands of actual Colonists.'[14] When Edward Stamp made an application for land near Langley in 1858, he clearly stated that he was not a land speculator,[15] reflecting an awareness that the government was inclined to favour permanent settlers with a commitment to the colony.

When Stamp began negotiating terms for land and timber in late 1859 for the Alberni sawmill project, the colonial government was unprepared for a venture of this magnitude. The discussions were complicated by the fact the land on Vancouver Island was still vested by a Royal Charter in the Hudson's Bay Company, although, as the Acting Colonial Secretary noted in early 1860 'that Charter if not already revoked, is probably on the point of being so, although no official information has been received to that effect.'[16] Moreover, the British government had not decided how land in the colony was to be alienated. Still, no problems were anticipated, and negotiations proceeded.

Stamp wanted a mill site on the coast with sufficient land for agricultural development, a licence for cutting timber, and the site declared a port of entry to facilitate the arrival and departure of ships.[17] Stamp wanted outright ownership of some land for the mill site and settlement, but he did not want to buy extensive timber lands. The costs of purchasing land that was essentially worthless after the timber was taken off seemed prohibitive. He just wanted cutting rights to timber on

Crown land. After discussion between Douglas and the Acting Colonial Secretary, Stamp was offered 2,000 acres at a price not exceeding £1 ($4.85) per acre. As for Stamp's request for an exclusive right to cut timber in specified areas, the government was not sympathetic: 'no exclusive right to cut timber in any particular locality can be granted to you, but ... a licence to cut timber in any part of the unoccupied Crown Lands can be granted to you under a royalty of ten pence per load.' Stamp was reminded that in return for these terms the government expected improvements of £7,500 within the year.[18]

While Stamp found most of the terms acceptable, he was unhappy with the timber terms. One issue was the cost: he noted that, in a well-timbered district, the royalty of 10 pence a load would amount to £15 per acre, and 'under such circumstances I respectfully beg to decline the licence.'[19] Moreover, without exclusive cutting rights in a particular area, there was concern that, after investment and improvements had been made, interlopers might arrive and begin logging and milling in the same area. In the end, Stamp was granted a timber-cutting licence that would include up to 15,000 acres, which it was believed would sustain the mill for twenty years, and he was charged a favourable rate per acre. The government retained ownership of the land.[20] This established an important principle: the separation of ownership of the land from timber-cutting privileges.

In 1865, when Stamp arranged for the construction of the milling operation on Burrard Inlet, events proceeded rapidly. In a letter of 17 May, Stamp presented his demands, and three days later the government replied: Burrard Inlet was to be declared a point of entry, the company could purchase a 100-acre mill site at $1.00 per acre, and Stamp would be allowed to import sawmill machinery without paying duty. Timber was to be allocated in the form of a lease. The company was to be allowed a lease of timberland consisting of up to 15,000 acres at a rent of $0.01 per acre per annum. In return the company had to build a mill capable of cutting 50,000 feet per day, and keep the mill in operation. If the mill closed for three months, it had to be justified to the Chief Commissioner of Land and Works and the Surveyor General.[21]

For the larger mills the twenty-one-year lease became the preferred form of tenure. In July 1865, S.P. Moody and associates requested a timber lease covering 5,000 acres 'on similar terms to the grant made to Captain Stamp,' assuring the government that, since 'the land on which this heavy timber grows is unfitted for agricultural purposes, it is therefore impossible for your memorialists to interfere in any way with the rights of actual or intending settlers.'[22] On 31 January 1866, S.P. Moody, James Van Bramer, and John Pickavent Cranford secured a lease covering 2,636 acres for the Moodyville operation. On 1 January 1870, Moody, Dietz, and Nelson increased their lease holdings by 11,410 acres.[23] Shorter-term leases and cutting licences were available for smaller operators that did not want the carrying costs of long-term leases. At Chemainus, on Vancouver Island, William P. Sayward secured a two-square-mile timber-cutting licence for a seven-year term at a cost of $0.02 per acre per annum in 1868.[24] In 1870, T.G. Askew secured a seven-year lease covering 519 acres to supply his Chemainus mill.[25]

The provincial government revisited timber-allocation procedures in the 1870s and 1880s. Most mills held timber leases in the 1870s, and the terms of the lease stipulated that the timber holder had to have and operate a sawmill. A select committee of the provincial legislature was created in 1875 to investigate concerns that the mill clause in leases made it too difficult for beginners with little capital to get involved in lumbering. The committee recommended a new form of temporary lease to facilitate the entry of small operators.[26] An Amendment to the Land Act in 1884 created a new form of timber allocation. Operators could obtain a four-year, non-transferable licence covering 100 acres. The cost was $10.00 rent per annum plus a royalty of $0.15 per tree. A fee of $0.20 per thousand board feet was also to be paid.

In 1886 the licences were divided into two categories: Special Licences, which met the terms specified above and applied to a particular tract of land, and General Licences, which allowed handloggers to take out timber from any unclaimed Crown land. Robert E. Cail suggests that these changes were brought in to make timber available to small, independent operators who did not own a sawmill.[27] However, there was no initial rush by logging contractors to take advantage of the new licence, suggesting that the pressure for the changes did not come from this quarter. In 1884, two licences were issued, but they were for the outlying regions of Cassiar and Lillooet. In 1885, six licences were issued, five of them in the coastal district, but only two went to men that ran logging camps, and the rest likely to handloggers. The lack of interest in the licences may have been due to the depressed economic times, but even after 1885 the popularity of licences, which were taken out by sawmill owners as well as independent loggers, increased slowly.[28] The licences also gave the government an opportunity to generate revenue for provincial coffers.

While both the needs of small operators and revenue were considerations, the timing of the amendment is best explained as a means to impose order and protect private property in the woods. In the buoyant times from 1882 to 1884, there was a log shortage, and as such logs were valuable: Hastings Saw Mill closed in February 1883 due to a shortage of logs. More people were going into the woods, and some were taking logs from the lands of legal leaseholders such as the Hastings operation. In December 1882, Hastings Saw Mill warned people about trespassing on their timber lands, and a month later they were compelled to run a notice in the newspaper: 'All persons are hereby cautioned against cutting or removing timber of any kind from all lands held by the Hastings Saw Mill Company, as parties doing so, without permission, will be prosecuted.'[29] The licencing system enabled the company and the government to keep tabs on loggers, remove the likelihood of misunderstandings, and protect the rights of leaseholders.

Leases remained the favoured form of tenure for the mills. In 1878, the Moodyville sawmill held 24,208 acres under lease, and Hastings held 18,559.[30] However, companies continued to log private land, although it is extremely difficult to trace the private holdings of the various mills. In 1883 and 1884, Moodyville Sawmills purchased 160 acres on Cortez Island; 320 acres on Malaspina Strait, 2 miles east of Grief Point; and 160 acres at the head of Malaspina Inlet. Royal City Planing Mills purchased 320 acres on Bowen Island, and land on Gambier Island.[31] The mills and logging contractors also arranged with owners of private land, often farmers, to take the logs off their property.

In the 1884 the Dominion government became directly involved in the coastal British Columbia lumber industry. Having settled the route for the railway, the federal government took up land in the province as compensation for their role in subsidizing the Canadian Pacific Railway (CPR). In the Settlement Act of 1884 the province turned over a twenty-mile-wide belt on either side of the line to the federal government. The Railway Belt included much fine timber in the Fraser Valley. In May 1885, the federal government announced regulations for the alienation of its timber. The charges of 30 cents per tree and 75 cents per thousand board feet averaged out to about 90 cents per thousand board feet, significantly higher than provincial dues, which averaged about 27.5 cents per thousand. The announcement came at an inopportune time as the lumber industry was in the doldrums. Half of the provincial sawmills were not operating: Dominion Sawmills and Moodyville Sawmills were closed, and Hastings was expected to cease production shortly. The industry sprang into action. A deputation of lumbermen visited

Victoria and gained provincial support in protesting the new levy, and a series of meetings were organized throughout the Lower Mainland. The New Westminster MP, J.A.R. Homer, and Senator Thomas R. McInnes pressed the provincial case in Ottawa. British Columbians were talking secession from Confederation. In June 1885, the federal government yielded, reducing timber rates in the Railway Belt west of Yale to the provincial level. The first two Dominion timber limits were taken up in December by W.D. Matheson, a logger, and W.C. McDougall, a logging contractor who had recently purchased the small Burr and Armstrong sawmill in Port Moody.[32] The Dominion timber largely fed the mills along the Fraser River, including the facilities at New Westminster. For the most part, Dominion timber regulations followed patterns established by the provincial government.

The railway deals of the early 1880s also led to the transfer of some 2 million acres of land on the east coast of Vancouver Island, constituting about one-third of the island, to private hands. The land was alienated by the provincial government in compensation for federal subsidies to railway builders. The land was transferred to the federal government, which promptly turned it over to private interests, headed by Robert Dunsmuir, in 1884, who were free to sell the timberland or use it as they saw fit. This alienation was done to facilitate the construction of the Esquimalt & Nanaimo (E&N) Railway, which was completed in 1887.[33]

The E&N Belt contained some of the best timber in the province, and as it was on privately owned land it was beyond the purview of governments with regard to cutting regulations, export laws, and stumpage and royalty payments. In the 1880s public concern over the grant focused on the mineral, not timber, rights given to Dunsmuir and his associates, but by the 1930s it was clear that the prerogatives of the holders of private timber in the E&N Belt seriously undermined the ability of the provincial government to manage the forest resources of the province. Historian Richard Rajala notes that 'the fundamental forestry question in the province prior to the Second World War concerned logging on private land, not the public forest.'[34]

There were minor revisions to the provincial Land Act in 1887, 1888, 1891, and 1896, and logging operators, including independent contractors, sawmill companies, and handloggers, continued to secure their logs from a variety of sources. Timber licences were still not particularly important. General Timber Licences allowed handloggers to log on Crown land and Special Timber Licences gave holders the right to log a specified area for one year (see table 4.1).

Table 4.1 Coastal Region Timber Licences, 1886–1897

Year	No. of special licences	No. of general licences
1886	15	19
1887	23	17
1888	34	62
1889	21	55
1890	17	17
1891	17	30
1892	28	12
1893	22	25
1894	19	8
1895	33	16
1896	30	16
1897	13	16

Source: Tabulated from British Columbia Forest Service, Original Licences, BCARS, B 4700, v. 1.

The acquisition of general licences for handloggers reflected the general ups and downs of the lumber economy, as there was no point taking out logs if there was no one to buy them in a depressed market.

The holding of special licences was more stable. Holders renewed their licence in anticipation of better markets in the future. Sawmill companies as well as independent loggers acquired special licences. For example, in 1888 the Hastings Saw Mill Company, as well as millmen such as W.P. Sayward, Albert Fader, Benjamin Springer, and G.F. Slater, held Special Timber Licences. Mills controlled even more licences than records show. Because legally a person or company could hold only one timber licence, companies took out licences under the names of others. Even many seemingly independent logging contractors who held licences were tied financially to the mills that guaranteed to take the log production from their site. Overall, in the 1890s timber licences were controlled by the mills, but they were still less important than leases in providing source of logs for production plants.

Privately owned land also provided logs, and mills acquired private timber land when possible. For example, according to newspaper estimates, in 1888 Maclaren and Ross purchased 2,500 acres of timber on the northern mainland coast from J. Prevost and W.J. Goepel for $25,000.[35] British Columbia had no restrictions on selling land that was chiefly valuable for its timber, and established mill owners were concerned that choice timberland was falling into the hands of ersatz homesteaders intent on controlling timber for speculation. Industry people wanted well-forested land set aside for the forest industry. Andrew Haslam, owner of Nanaimo Saw Mills and representative of coastal sawmill owners, wrote to the government concerning the fears of lumber operators: 'we believe it would be an act of wisdom and justice to prohibit the pre-emption of any land covered by lease that has an excess of six thousand feet per acre of milling timber.' But until 1905 timberlands continued to be sold.[36]

In the late 1880s the Esquimalt & Nanaimo Railway began selling its privately held timberland. In January 1889, John Humbird of the Victoria Lumber & Manufacturing Company negotiated a deal with the railway, which allowed Humbird to purchase 100,000 acres of timberland on the east coast of Vancouver Island between the Chemainus River and Comox for $5 an acre. By 1895 the company had selected 88,393 acres, divided into forty-one separate parcels.[37] The 1891 cut on provincially controlled land was estimated roughly as 33.8 million board feet from timber leases, 16.0 million from licensed timber limits, 16.1 million from Crown lands, and 17.2 million from private property.[38] Logging operators also cut timber from limits controlled by the federal government in the Railway Belt. Almost all of these limits were taken up by operators from New Westminster and the Fraser Valley.[39]

As well as timber-allocation procedures, clear definitions of the boundaries of private property, both geographically and legally, were absolutely essential for commercial lumbering. Capitalists needed security of ownership to reap the rewards of their investments. In the colonial period, ad hoc alienation of land, vague oral commitments by government officials, and delays in official surveys caused consternation for lumbermen. On 2 February 1860, Thomas Donahue of New Westminster complained to the government: about nine months previously he had erected a sawmill on ten acres which he owned on the understanding that in the future he would have access to as much timberland as necessary to feed his mill. In January 1860, several parties pre-empted land on the boundaries of his property and expected him to pay for trees cut on their pre-emptions. Donahue was furious. He felt that the government had reneged on a commitment and he was especially angry because he had spent money and energy building his business while the pre-emptors had done nothing and expected to profit from his work. The government was sympathetic but unable to act.[40]

At Alberni, the holdings of Anderson & Co. were surveyed in early 1862, but Stamp was unsatisfied with the official map and demanded corrections.[41] The boundaries and status of the holdings at Alberni were still in dispute in 1864, and the company was alarmed that the government was planning to sell land at Alberni, claiming that the advertised land encroached on the company's land claims. The company was also concerned that new settlers would be free-riders, taking advantage unfairly of their costly improvements made in the area. Negotiations ensued and the company abandoned a number of its claims for secure tenure. In 1871, however, Anderson & Co. was still seeking confirmation that the deal of 1865 was valid.[42] In January 1870, S.P. Moody & Co. complained that pre-emptors had taken up land on their timber lease. They had a logging camp on the lease and had spent $2,000 on roads; moreover, 'during the last two years we have expended over One Thousand Dollars, in extinguishing and keeping under fires started by Indians, and malicious persons, for the purpose of destroying timber ...'[43] In the spring of 1871, a surveyor laying out a Crown grant on Burrard Inlet noted that there was conflict with the holdings of Hastings Saw Mill: 'is this not another blunder in the numbering, as the Hastings Mill Co.'s land is numbered the same?'[44]

The most celebrated conflict over property rights in the colonial lumber industry was a result of sloppy government procedures. The Military Reserve on English Bay contained high-quality spar timber. Governor Seymour orally gave Edward Stamp rights to the timber, and Jeremiah Rogers, who had logged for Stamp since the establishment of the operation at Alberni, set up a camp and began taking out spars. Like many others, Rogers had a falling out with Stamp and began sending timber to the Moodyville mill on the other side of Burrard Inlet, as well as shipping out spars on his own account. In 1867, Rogers applied for a formal cutting licence, and the government prepared an indenture. At this point, Stamp intervened, stating that he had the rights to the timber, and demanded that Rogers vacate the reserve. Stamp accused Rogers of speculation and blackmail. In late April, Stamp forced a court case, and Rogers and Jonathan Miller, another logger on the reserve, were charged with trespass. The case was dismissed. Seymour appealed to the parties to settle the issue amicably, but was unsuccessful.

The government then decided to stop all logging on the reserve; on 26 June 1868, Rogers was ordered to stop logging, as he was 'logging without authority.' Rogers, supported by many at Burrard Inlet and New Westminster, believed that he was acting honourably because he had an oral commitment from the government to log the reserve. Indeed, both Stamp and Rogers had oral commitments. In July, another court case was launched against Rogers and Miller. There was outrage and criticism of the government, and especially Edward Stamp, who was seen as orchestrating the whole affair. Stamp was portrayed as a 'snake in the grass,' intent on undermining the hard work of locals such as Rogers and Miller. John Robson, newspaper editor and foe of Stamp, wrote: 'Not content with the enormous timber reserves he has already succeeded in monopolizing, he is intent on placing his paw on every acre of timber land within reach,' adding that Stamp's 'selfish machinations' were 'inimical to the best interests of the Colony.'[45] On 31 July 1868, a public meeting was held in New Westminster to support Rogers and Miller, and a resolution was drafted and sent to Governor Seymour. Finally, in November 1868, the issue was settled: Rogers was given a five-year lease on the Naval Reserve.[46]

With general population growth and increased logging activity, disputes over land and timber rights intensified after 1886. In 1890 Alexander Smith, a Chilliwack farmer, protested the granting of a timber lease to Knight Brothers, claiming that the timber land was rightfully his, and a group of thirty-one settlers in the Seymour Creek area protested a timber-licence application by John W. Taylor, arguing that they should have the right to do the logging. Pre-emptors in the Squamish area demanded that a number of timber licences not be renewed in their area, and Fred T. Lazenby, representing twenty settlers from Port Hammond, was furious that the provincial government had granted a lease to Brunette Saw Mills in his area: he claimed that after the lease was issued many local settlers moved, many to the United States.[47] Lumbermen, for their part, were concerned that granting mineral licences on timber leases destroyed too much valuable timber: 'the government should exact a bond from the prospector sufficient to recoup the mill owner for losses he might sustain through the destruction of timber by mining operations.'[48] Lumbermen faced other problems. In 1891 a boom of logs taken out by a company of Maple Ridge farmers was moored on a river where it emptied into the Pitt. The boom was cut, and the logs dispersed. The cutting of the boom was legal because no boom was allowed to obstruct the whole width of any stream.[49] In 1892 the Burnaby Lake Lumber Company was temporarily forced to close its logging camp because of an injunction initiated by the New Westminster Water Works Company, which claimed that logs being run down the stream to Burnaby Lake struck their pipes.[50]

Lumbermen also had differences with each other. In 1890 the Brunette Saw Mill Company complained that Isaac Hayden, who was taking out logs for the Webster & Edmonds sawmill, was logging their timber at Storm Bay on Sechelt Inlet. They demanded that Hayden's logs be seized for trespassing. To the north at Port Neville, D.C. Esson protested the granting of a lease to the Brunette sawmill, claiming that he had 'already done considerable work on this land in constructing roads and opening up the claim for work.' Esson claimed to have a contract with Leamy & Kyle's False Creek sawmill operation. Then Leamy & Kyle protested both the Brunette and Esson claims to the timber. They had been refused the lease in 1888 because a portion of it was owned by T.D. Merrill, but in the meantime they had purchased the rights from Merrill. Thus, in 1890 they claimed that the government in fairness should give them the lease.[51] In 1888 a logger on the North Arm of the Fraser pointed a gun at a competitor in a dispute over timber-cutting rights.[52]

The pressure on the land and resources in the coastal region by water, farming, mining, and lumber interests was increasingly evident by 1890. Property rights regarding logs in transit also became of greater concern. With increasing numbers of log booms plying the Fraser and the inside passage, it was difficult to identify the owners of logs when a boom broke while on the water. In 1890, Andrew Haslam, lumberman and member of the provincial legislature from Nanaimo, ushered a bill through the House which provided that each logging operator had to register a particular brand, which he had to use to mark his logs, in order to establish ownership.[53]

An issue which united British Columbians was the devastation caused by forest fires. Fire was perceived as the greatest threat to the forests and the lumber industry, but preventing fires was not a simple matter. In the colonial period, forest fires threatened towns, but as John Robson commented in 1868, 'now the forest has retreated before the woodsman's axe to a safe distance.' The damage to timber and threats to logging camps, however, continued, and in almost every summer fire took its toll. In 1868, 1870, 1872, and 1875 Burrard Inlet was full of smoke from a number of raging fires. In 1882 a fire was located at Jeremiah Rogers's old spar camp, moving towards Rosehill farm; another fire on Howe Sound was so near a logging camp that half of the crew were sent away with the cattle; and yet another forest fire, at Jervis Inlet, forced loggers to seek refuge by fleeing to the seashore. In 1883 a New Westminster correspondent noted that bush fires had 'spread over the whole country. Everywhere one goes there is smoke, smoke, smoke. On the Granville road, the Port Moody road, down the North and South Arms, across the river, up both sides of the Fraser as far as Yale, and up the northwest coast, all is fire and smoke.' A fire in 1885 burned out a Moodyville Sawmill Company logging camp and inflicted much damage on a Hastings camp. In 1887, while fires raged near False Creek, a fire started in McDougall's logging camp at Mud Bay, destroying timber held by Royal City Planing Mills. The years 1889 and 1891 were also bad for fire, and in 1895 fire destroyed the Haslam camp on Valdez Island: 'The fire is running through the camp and has burned bridges and skid roads, so that they cannot be used. They had an engine for hauling the logs on to the roads and had hard work to save it.'[54]

The Bush Fire Act of 1874 provided heavy penalties for starting a forest fire, but because it came into force only if two-thirds of the settlers in a district signed a petition, it was problematic because there was no census to establish what number constituted two-thirds in a given area and it demanded initiative at the local level. Moreover, the law was not enforced. In 1889 there was a call for a vigorous enforcement of the law, encouragement to citizens to turn in fire starters, and a suggestion to establish police or detective surveillance during the most dangerous months for fires. In 1890, Henry Croft, Member of the Legislative Assembly from Cowichan and lumberman, ushered a new Bush Fire Act through the legislature. One clause stipulated that half of the fines and penalties imposed on persons who contravened the act would go to the government and the other half to the informant. Enforcement, however, continued to be lax, and the new law did little to curtail the scourge of forest fires.[55] The increased use of steam-powered machinery in the woods after 1890 exacerbated the problem, due to sparks from railway locomotives and steam donkeys.

The provincial government was also drawn into the thorny issue of scaling. Scaling or measuring timber was extremely important to logging operators, millmen, and the government. The accuracy of the scale of a log in board feet and the rules or standards used to calculate the amount of lumber in a log directly influenced the loggers' income, the mills' cost for logs, and how much the provincial government could claim in revenue. Logs were scaled when they arrived at the mills and it was done by mill employees, who depended on the mill owners for their wages. The logging operators became increasingly distrustful of the system which they felt was weighted in favour of the millmen. By 1892 logging operators were calling for the appointment of a government scaler to oversee the system and ensure that the calculated scales were fair.[56]

The government promised to address the matter but was slow to act. One logger vented his frustration in December 1893: 'All we ask is a fair deal. The mill companies have had it all their own way long enough and we don't propose to stand it any longer ... As far as I can see the only qualification required to scale for the mill companies is to cut the scale down as much as possible.'[57] In 1894 an act established a system of government scalers, and in the fall candidates were being tested by a board of examiners. Official log scalers were appointed to supervise areas of the province. The question of the standard by which logs were scaled was also addressed. The Chief Commissioner of Lands and Works struck a committee in 1895 to devise an official scale that was acceptable to both loggers and millmen, and they established the British Columbia Log Scale. By 1898 the millmen were complaining about the new system, and the whole issue of scaling continued to fester.[58]

E.H. Heaps remembered that 'it was more or less a question of argument between the logger and millmen as to what the scale of a boom was but the general result was that the mills acquired a supply of logs cheaper than $5.'[59] In 1902 a joint committee of loggers and millmen developed new scaling regulations that were enshrined in provincial legislation, but scalers, who were appointed and monitored by the government, were still employees of the mill company, thus putting the scalers under pressure to calculate scales acceptable to the mill owners. Standards for grading logs and the formula used to calculate the scale were also seen differently by mill owners and loggers. Legislation in 1906 created a system of independent government scalers for the coastal region, but boss loggers continued to claim that the whole system favoured the mills, to the financial detriment of the loggers.[60]

Governments were very involved in creating a hospitable environment for commercial forestry and, more specifically, in devising acceptable timber-allocation procedures. However, revenue from forest resources was minimal. In the last half of the nineteenth century, the forest industry was too tentative to generate revenue for government coffers, and governments were content with the economic activity generated by the forest industry. During the depression of the 1890s, the provincial government practised sympathetic management, allowing companies to delay royalty and licence payments that were owed to the government. In commenting on the low revenue from the forest industry during the Budget Speech debate in February 1894, John Turner justified the government's position. He argued that, if the government pressed the industry, it would lead to mill closures, unemployment, and possibly even the crippling of the industry. He was confident that the charges would be paid when the trade improved.[61] Only in extreme circumstances did the government act against companies: in 1895, during a short upturn in the lumber economy, the government seized the Moodyville Sawmill Company, closing the mill, due to substantial arrears in timber payments.[62]

While the provincial government practised sympathetic management, the federal government provided no management. The position of timber inspector, within the Department of Lands and Works, was created in 1888, but with no administrative staff, he merely watched and monitored, to the best of his ability, the operations of the industry. Lack of personnel and commitment meant that loggers and mill companies acted largely on their own initiative. In 1897 the federal government launched an investigation into the management of the Dominion timber office in British Columbia. Although the commission was called by a new Liberal government in Ottawa, it exposed many problems, forcing the resignation of Thomas Sheriff Higginson, the Crown Timber Agent in New Westminster since 1884. The comments of the commissioner, Archer Martin, were very damning regarding Higginson and the department: 'I am forced to report that his [Higginson's] neglect of duty has been so gross as to afford only two solutions for his conduct: either that he had a callous indifference to the preservation of the property entrusted to his care, or that he had been induced to connive at the irregularities which flourished under his eyes.'[63]

By 1900 the provincial government had a developed forest policy. There was a commitment to industrial growth, and the aim was to allocate timber only to companies that had operational milling facilities. The bigger the milling company, the better. The provincial government, following British traditions, retained ownership of the forest land, with the notable exception of the Railway Belts, giving the state a direct interest in the forests and a continued role in shaping logging practices. The state expected revenue from the forest resource, but the economy of the nineteenth century undermined such hopes. The economy through the 1880s was weak, and the rapid expansion from 1885 to 1892, which finally held out the possibility for revenue, came to a crashing halt with the depression of the 1890s. The revival of trade and the rapid growth of the industry after 1898 challenged the existing policy structures, and new rules were created for a new environment.

5

Critics

In the 1890s, the language and ideas of the continental conservation movement arrived in British Columbia. R.E. Gosnell, in his 1897 British Columbia yearbook, comments that 'forestry, as an economic science, is one to which very little attention has been paid in the Province.' But he adds that this was changing and that soon 'it will be possible ... to regard forestry as a branch of agriculture, and to speak of a wood crop, to reap which it will not be necessary to wait a lifetime.' Gosnell wanted to control forest fires, plant trees, minimize wasteful logging practices which left too much wood in the forests, and encourage pulp-mill development. Pulp mills were more efficient than sawmills because they used smaller and younger trees, and thus timber limits could be logged clean. He also believed that a scientific understanding of the biology of coastal trees and the economics of markets for local wood products would allow for rational use of the forest, a notion that would later be called 'sustained-yield management': 'It is not improbable that a careful study of the economics of forestry would develop a system in connection with tree growth and forest clearing that would ultimately equalize supply and demand and render the absolute destruction of wood as at present unnecessary.'[1]

A few years later, J.R. Anderson, British Columbia's Deputy Minister of Agriculture, argued that, if the province did not quickly begin to apply progressive forest-management techniques, based on sound science and economics, the future of the provincial forests and forest industry would be bleak. Using estimates of the total amount of timber in the province and guessing at the size of the timber cut each year, he predicted that 'the present limits would be worked out in twenty years.' But recognizing the weakness of his data, he also offered a more conservative view, drawing on a different estimate of provincial timber holdings, in which the forest would be cut in sixty years. Both predictions were predicated on the belief that log production would treble in the near future.

To cope with the increased cut, and to maintain the industry and a productive forest, Anderson felt that more progressive forest management was necessary. The main culprits in the 'continued destruction of our forests' were 'wasteful methods, carelessness, and above all, forest fires.' He estimated that bush fires consumed fully 50 per cent of the timber. In his discussion he drew on evidence that described the depletion of forests and the decline of the lumber industry in the Midwestern United States that had occurred a generation earlier. Anderson advocated a forest-ranger system like the one that operated in Germany. He was also keen to point out that the practice of forest conservation did not spring from altruism but was rather based on sound business principles. Anderson's argument was presented at a meeting of the Canadian Forestry Association in Ottawa in March 1901.[2]

The push for more rational, progressive forestry management spread throughout North America in the last quarter of the nineteenth century, uniting lumbermen, professional foresters, and governments, and leading to laws and regulations to conserve forests and sustain the forest industry. Already in the early 1880s forest conservationists in Ontario and Quebec were working together with Americans to perfect organizations to promote forest conservation.[3] At its heart, the conservation movement was about applying science and technology to ensure rational, profitable harvesting of timber. As Samuel P. Hays writes, 'its essence was rational planning to promote efficient development and use of all natural resources.'[4]

It is worth emphasizing that this forest conservation movement was not against business or logging – it was not preservationist – and though it was driven by a belief in the science of the day, it was predicated on business, scientific experts, and governments working together. Bernhard Eduard Fernow, a key figure in the movement in the United States and Canada, was at pains in a letter from the early 1890s to distance the movement from fringe ideas and people:

> Now as to the 'forestry craze.' When it is divested of its crude, more or less unintelligent interpretation by uninformed friends – such as every reform movement must count in its ranks – it is not to stop lumbering or to introduce European management or to advocate anything irrational. On

> the contrary, its object and ultimate goal is to advance more rational use and treatment of our forest resources; to aid in insuring continuous supplies for the lumber business; to stop only wasteful practices, especially destruction by fire ...[5]

Despite the extent of its timbered areas, British Columbia was slow to participate in the larger, continental movement, and Anderson's journey to Ottawa in 1901 was evidence that the province was finally being integrated into this intellectual, political, and economic milieu.

In many ways Anderson was a good representative of the conservation age in its broadest outline. While he was neither a forestry expert like Fernow nor a businessman, his activities captured the breadth and optimism of conservationists at the turn of the century in British Columbia. A civil servant who had joined the Department of Agriculture in 1891, he recognized the role that the state could play in managing the resource and he knew the importance of information in decision making and changing long-standing attitudes. His scientific approach reflected a new direction in forestry management in British Columbia. In the 1890s, without financial or staffing support, he began to gather data on the economic value of the provincial forest, its extent, and its rate of depletion. As early as 1892 he recommended the 'reservation of a tract of virgin forest as a sanctuary for the preservation of at least a remnant of our primeval forests,' but to his dismay the first provincial park, Strathcona, was not created until 1910.[6] Anderson also read about forest practices and procedures in Europe, the United States, and central Canada, drawing on these to formulate strategies for British Columbia, thus locating the province in the broader intellectual world and rooting his discussion in the language of statistics, comparisons, case studies, and science. Anderson also made links between logging practices and the survival of agriculture and the fishing industry, as well as importance of proper forest management for the preservation of a water supply for human consumption.

Although the language of the conservationists gained acceptance in the first decade of the century in British Columbia, it was not an easy, smooth transition. Scientifically based experts in universities and government bureaucracies clashed with politicians and businessmen. Many were sceptical of the ideas of the conservationists. Even a knowledgeable person such as R.J. Skinner, the provincial Timber Inspector, was leery of the conservationist's agenda in the first years of the twentieth century, and his beliefs likely reflected those of many lumbermen.

Skinner preferred to rely on market forces and the decisions of lumbermen, instead of experts and bureaucrats, to dictate logging practices and forest policy.[7] Indeed, Skinner's concerns provide a clue to why conservationist principles arrived late in British Columbia and also why they were eventually accepted. Changing market forces after 1897 made coastal timber more valuable, and because there was money to be made, companies saw the merits of logging more efficiently, removing more wood from the forest, and adhering to conservationist principles. It became clear that conservation ideals were not a threat to business. Conservation did not merely serve the industry; it also meshed with market conditions in British Columbia.

But we are getting ahead of the story. New notions regarding forest policy and practices emerged and came to dominate discussions in the first decade of the century; however, this chapter looks at the critiques of government policy, of logging practices, and of the structure of the industry in the last half of the nineteenth century. This aspect of the history of the forest industry in the province is less well known. Most of the concerns raised by the turn-of-the-century conservationists had been discussed in British Columbia since the 1860s: the need to curtail forest fires; the wasteful, careless practices of logging operators; the depletion of the forests in other parts of North America; the weaknesses of government monitoring; and fears about the long-term viability of a lumber economy.

But, importantly, the later conservation movement was not a coming together of the ideas that circulated in the province from the 1860s through the 1890s, nor was it a carried on by the same people. Rather, it was a shift, a different way of looking at the world, that supplanted the older critical tradition. Whereas the later conservationists argued for a scientific approach, rooted in objective forms of knowledge that attempted to remove human and social considerations, the nineteenth-century critique was embedded in a social and political perspective. In the earlier view, connections among monopoly capitalists, abuses of government regulations, wasteful logging practices, politicians, and the destruction of the forests, with little return to the people of the province, were rolled together in a dissenting vision. In this vision, evil forestry and political practices stunted economic growth; caused farmers, workers, and independent small businessmen to suffer; undermined democracy; and perverted a sense of community. The knowledge that informed the critique was not scientific; it came from hearsay, local sources, and personal experiences, and was interpreted through the prism of an ideal community made up of small-scale producers working together for the benefit of the whole. Human beings, social and economic institutions, and the environment were all important and interconnected. In this view, resource management was inherently a political issue.

The movers and shakers in the later conservation movement portrayed themselves as beyond politics. Samuel P. Hays captures this aspect of the movement: 'The political implications of conservation, it is particularly important to observe, grew out of the political implications of applied science rather than from conflict over the distribution of wealth ... Since resource matters were basically technical in nature, conservationists argued, technicians, rather than legislators, should deal with them.'[8] In reality, though, the principles of the later conservation movement in British Columbia supported large, well-capitalized companies over smaller operations. These large business concerns had the technology to clear-cut wide swaths of forest in an efficient manner, which, it was argued, was the best site preparation for managing a new crop of trees. Large companies also had the financial resources to weather the ups and downs of the economy and thus allow for stability and long-term forest management, and the wherewithal to hire forestry experts to ensure the rational use of the forests. These features of scientific forestry, of course, contributed to the marriage between big companies and the turn-of-the-century conservationists. In the earlier critique big companies were the enemy.

The critical perspective on the nineteenth-century provincial lumber economy was in many ways an attack on the type of capitalism being created in British Columbia. The argument was that big companies controlled politicians and timberland, thwarting the creation of a marketplace where a wide array of middling interests controlled capital and production, and interacted from positions of relative equality in the interests of the local community and the forestry resource. This dissenters' critique was an undercurrent in nineteenth-century British Columbia. Sawmill owners and logging-camp operators were largely unfettered, with politicians doing as much as possible to encourage large-scale capital investment. John Robson, newspaper editor, unabashed booster of the lumber interests, and later premier of the province, argued in 1865 that the government should do as much as possible to encourage Edward Stamp and the lumber industry generally: 'It is hardly necessary for us to express the hope that the Government will meet Capt. Stamp, and indeed any other parties proposing to engage in a similar undertaking, in the most liberal manner.'[9] A good proportion of British

Columbians would continue to endorse this perspective. However, the dissenters were evidence of an alternative perspective on development in the province, extending the boundaries of political discourse and shaping the culture of British Columbia politics.

Much of what follows comes from the pages of *The Mainland Guardian*, a New Westminster newspaper edited by J.K. Suter from 1869 to 1889. Suter was the voice of dissent during the 1870s and 1880s, and his words provide an entree into the alternative social and political vision. At first blush it is tempting to dismiss Suter as a crank: his editorial comments were vitriolic, punctuated with personal attacks; his arguments were often based on rumour and poorly substantiated; and the causes and politicians he supported usually lost. However, a couple of things suggest that Suter spoke for a sizeable constituency.

First, Suter was located in New Westminster, and he wrote for and represented the views of the rural population in the Fraser Valley. Farmers were often critical of the lumber industry. As we have seen, farmers clashed with lumbermen over tariff policy and the purchasing strategies of the mill companies and logging camps because they did not buy enough locally. Farmers also frequently criticized the land policy of the government, which they believed favoured the lumber operators. For example, in 1884 an outraged farmer wrote to the provincial government protesting a timber lease granted to a New Westminster sawmill company at Mud Bay. He had been farming in the area for four years and had made sufficient progress so that he was supporting his family. Representing some forty farmers, he protested the turning over of timber to a mill: 'We think the Government does not intend to treat us in such a shabby manner leaving us without any timber for fire wood, building or fencing ... All this section of the country is good land and will be cleared and improved. We want the timber on the land to help us improve the same. This is not an isolated place only fit to cut timber, but good land and in a thriving settlement.'[10] Fraser Valley farmers, of course, were not completely hostile to the lumber interests. Farmers ran logging operations, worked in logging camps, and sold farm produce to lumber companies. In the 1870s, William Ladner raised 'the heavy seven-foot girthed oxen demanded in logging' camps.[11] But the farming community occupied a particular niche in the economy and was keen to defend is material interests.

The second aspect of Suter's dissenting voice which suggests that he spoke for a broad constituency is that, in representing material and class interests, he tapped into a deep intellectual tradition in nineteenth-century Canadian politics. In other words, Suter was not a loose cannon with some idiosyncratic ideas, but rather reflected an established, recognizable strain of political thought in Canadian society. The terms 'radical liberalism,' 'republicanism,' and later 'populism' and 'labourism' capture aspects of this intellectual world view. Peter J. Smith, in a study of ideologies in Canada at Confederation, notes that a fundamental cleavage separated petty agrarian producers and large-scale commercial interests. The small producers celebrated virtue and independence in a decentralized political system, associating corruption, privilege, and centralized power with the rising commercial and financial interests.[12] As the century progressed, many Canadian workers adopted a similar perspective on capitalism.

British Columbia was not immune to these ideas. In a study of the Vancouver union movement between 1889 and 1910, Mark Leier describes the labourist doctrine espoused by the craft workers, a doctrine connected to agrarian populism. In this view the world was divided into producers and parasites; farmers, artisans, small manufacturers, and small proprietors were ranged against the great monopolists.[13] This sentiment was expressed at a meeting of the Workingmen's Political Club in New Westminster in 1893. In protesting the government's land policy, it was argued that all land was either in the hands of speculators or covered by timber leases, leaving no opportunities for ambitious, hardworking men with little capital.[14] This ideology of dissent, then, was far-reaching, with exponents in the farming, working-class, and small-business communities. It celebrated a democratic society built on the labour of small producers and was deeply critical of monopolists who controlled corrupt politicians to loot the land at the expense of the people. By 1870, J.K. Suter had already concluded that the lumber interests of Burrard Inlet were tightening their stranglehold on the economy and the government.

Since Suter's forestry critique and his views on political economy were all of a piece, a broad perspective is necessary. Suter believed that farmers were the backbone of a healthy society, and that large-scale capitalists, with the ear and support of the provincial government, offered little to society but reaped all the rewards. In 1874 Suter contrasted the activities of lumbermen and mineowners with the fate of farmers, who 'really pay the salaries of the public servants':

> [Farmers] toil day after day in making this country rich, by clearing and cultivating the lands. They live in out of the way places and are denied the

> advantage of social communion with their fellow citizens, very often. They are exposed to all kinds of discomforts, and are frequently short of the means to purchase common necessaries. In the course of a long and frugal life they may hope to accumulate sufficient to leave their children in comparative comfort. As they are the real workers, and those who are making the country valuable, of course they are those who are most cared for by the Government. Not at all; they appear to be particularly those that the Government delights to harass and squeeze the last dollar out of. We make splendid wagon roads for wealthy traders, and ask these unfortunate tillers of the soil, who never use them, to keep them in repair; the settlers are taxed for roads and a variety of other specious reasons, and if they don't stump up, they are summoned and the taxes doubled with unnecessary law expenses. They have no roads in the vicinity of their farms, and have occasionally been unable to get their produce out. But, the Government officials are very assiduous in helping them when they want to do any business. Oh! Are they. These very ill used tillers of the soil, are met at the very threshold with demands for outrageous fees for all kinds of things, and when by dint of the most exhausting labor they have managed to get up their house and a certain amount of fencing; they apply for their certificates of improvement, the most vexatious delay is the common experience.[15]

While farmers laboured at productive work, the moneyed interests worked with politicians and government officials to line their pockets. Suter slammed the consortium of politicians and businessmen, including lumbermen, that was put together in the early 1870s, seemingly with insider information, to develop mining interests on Texada Island and on Silver Peak near Hope. Questions regarding the propriety of the link between business and politicians led to an investigation by a royal commission. The timber-lease system also outraged Suter, who argued that the province's most valuable land was falling into the hands of a few prominent lumbermen. The renewal of the Hastings Saw Mill lease in the 1880s, which involved closed-door negotiations with the government and the Canadian Pacific Railway, made Suter dyspeptic. The renewal of the lease, which meant 'granting to someone, the right to cut timber on the most valuable timber lands in the Province ... would be simple public robbery, and that of the most impudent kind.'[16]

Suter's alternative to the tying up of large chunks of timberland in the hands of a few was to open the land up to industrious settlers and to encourage the development of a number of smaller, competitive sawmill operations: 'The timber land held by them unworked, would have better paid the country had it been in the hands of rival Mill owners, who employed each a number of workmen.'[17] On another occasion he wrote: 'there are numbers of our own citizens with small capital, who would gladly enter into the manufacture of lumber, if they were permitted; but the late government took care to discourage all but their special favorites, and so maintained an unjust monopoly.'[18] In 1888 he advocated the letting of leases to small holders, and they would be controlled by government regulations to preserve young trees and ensure the regeneration of the forest.[19]

The meagre financial return captured by the province for the privilege of cutting public timber was another component of the dissenters' critique. Since the province's timber belonged to the Crown, the people as a whole should benefit from the cutting of the peoples' resource. Calls for higher lease, rental, and cutting fees were a constant refrain in the nineteenth century. This money would be used for the general development of the province. The forests should not be exploited for the benefit of a few capitalists but for the advantage of the community at large. A letter to the editor in 1872 put the case bluntly: 'What is wanted now to protect the public interests is a Stumpage law. The lumbermen have had the privilege long enough of cutting down our finest timber without paying a cent for stumpage, and so defrauding the public of their just rights. The people cannot afford to be robbed any longer. The produce of a stumpage law is very much wanted for schools, roads, etc.'[20] In 1876, Suter argued that, if the terms of existing leases could not legally be changed, then increased taxes should be imposed on the leased land, adding that handloggers should be charged for the right to take timber from the public domain.[21] In 1884 he argued that the renewal of the Hastings Saw Mill lease at only $0.01 per acre per annum was a travesty, suggesting that the land be thrown open to public auction, where Suter estimated it would garner some $1.2 million for the provincial treasury. The money could be used for schools and roads.[22] In 1889, the *News-Advertiser*, a Vancouver newspaper, opined that in British Columbia the Commissioner of Lands 'practically has a shingle out notifying the world that here there are "the finest timber lands on tap at all times" and the price charged is so small that the purchaser cannot fail to find it a profitable speculation to acquire them.'[23]

There was scepticism about the extent of the contribution of the lumber industry to the development of the community. In 1860, when arrangements were being made between Edward Stamp and the colonial government regarding the Alberni operation, Alfred Waddington spoke against the proposal, contending that 'all profits would go to England; everything required, from nails down to shoestrings, would come from the mother country. Victoria, would not be benefitted.'[24] In 1874, Suter questioned the contribution of the lumber industry to the local economy: 'The most extensive of our lumber manufacturing firms absolutely consume as little of our home produce as possible. They do not patronize our dealers and importers; they have all their necessaries brought up in ships from an alien country, free of freight, thereby evading even the little amount of coasting trade that the transmission of these goods would create.' He was also cynical about the contribution of the lumber industry work force to the development of provincial society: 'The very labor they hire does no good to the country, as it is of that wandering restless class that rarely settle anywhere.'[25]

Arguments about the lack of revenue generated for the provincial treasury and the questionable contributions of mill owners to the economy were situated in a larger concern that saw lumbering as merely a temporary, transitory industry. British Columbians were well aware of the fate of other local North American lumber economies. John Robson was forced to address the question of the longevity of the lumber industry already in 1868:

> In regard to the duration of the lumber interests, it is impossible to see all the future; but we may say the development of that source of wealth has only just commenced; ... we see no reason why it may not last for ages. As the forests near the mills melt away, farther and farther into the interior will our regiment of hardy axemen penetrate, until the mountain slopes, the bays and inlets, the rivers and creeks will all be placed under tribute to the mills; and, peering away into the future, when all these shall have been exhausted the smaller growth which was first passed by will have become ripe for the lumberman's axe.[26]

Others were less sanguine. In 1874, Suter cited the depletion of the forests in the United States as a warning to British Columbians; in April 1884, his newspaper reprinted an article from a New York state publication which argued that the tariff against British Columbia lumber was accelerating the destruction of the forests in the U.S. Pacific Northwest, quickly bringing them to a condition that had been created in the eastern United States. In July 1884 he claimed that British Columbians, who believed that the provincial forests were inexhaustible, were living in a fool's paradise, as there was little timber left in the province and it was disappearing rapidly.[27] Another commentator made similar points in 1880: 'While not many years ago the great problem was how to get rid of the noble forest that covered the land, the problem now is how to renew it. People in this Province might do well to profit by experience on the other side of the continent in this respect.'[28]

In the late 1880s, during an increase in logging activity, statistical data on forestry in the United States were used to illustrate the 'ruthless slaughter' of the forests in the U.S. Midwest: 'From the standpoint of to-day, it is indeed difficult to realize that the vast resources of British Columbia can soon be exhausted; but the fact will sooner or later be proven beyond any question of doubt. The same process of destruction which levelled the magnificent pine forests of New England, New York and Pennsylvania, and has swept over the timber lands of Michigan, Wisconsin, and Minnesota, will find its way to our fair Province and follow the same course.'[29]

The belief among dissenters that the forests were limited, and the industry temporary, prompted three avenues of thought. First, as we have seen, if the lumber industry was to be short-lived, then the government should exact as much revenue as possible from the destruction of the forests. Second, if the companies engaged in lumbering were not financially able to return to the people of the province a fair compensation for the timber logged, then the forests should not be worked. J.K. Suter began offering this line of argument in 1870:

> Some would appear to think that if we could only keep the two establishments going at the Inlet we should be doing extremely well; the employment of two or three hundred men to whom we could sell our groceries and dry goods, who are sending away our valuable timber for a small percentage over cost of manufacture. If this is the height of our ambition, we are to be pitied; a few years will see this splendid source of wealth exhausted, and those now engaged in sending it away, will be the first to leave us in our treeless poverty. Valuable lumber such as we are now so anxious to get rid of for a song, will be only be obtainable far in the interior; and we shall pay the trifling profits we are now obtaining back again tenfold with a fearful interest, when we have roused ourselves to a sense of necessity for action.[30]

He developed the argument in 1876. Anticipating the construction of a railway linking the Pacific coast with central Canada and an increase in the population of the province to more than 100,000, he foresaw that there then would be a great demand for lumber in British Columbia. But the people of the province would be in a sad state; having exported their timber to benefit only a few men making rapid fortunes, they would be forced to import lumber at extremely high costs: 'We are now giving away the golden eggs of the goose, and when we begin to understand their value, we shall find that she has ceased to lay.' He offered the same argument in 1884, but in this instance he was anticipating the day when the province's population reached one million.[31]

The third response to the limits of provincial forests was to encourage efficient use of the resource. Logging operations were criticized for leaving too much merchantable timber in the woods. In order to save labour costs, logging bosses sent out one chopper instead of two, and 'the one chopper climbs up our noble trees twenty feet, before he begins his labor of chopping. The result is, that the best part of the timber is left to rot on the ground!'[32] Loggers, under the guise of pre-emptors, took up well-timbered land under false pretenses, logged it, and then abandoned it: 'Now that quarter section is rendered useless, as no settler would take it up to clear it of stumps and branches, while he can get Land where no hand loggers have disturbed the timber, or left a chaos of worthless stumps and refuse branches to be cleared away. This we conceive to be a prostitution of the Pre-Emption Act ...'[33]

The practice of logging only the choicest timber also generated anger. Operators took leases and licences on only the pockets of the forest that contained high-grade trees, making it economically unfeasible to go in afterwards to take out smaller trees and less valuable species: 'active steps should be taken at once to utilize every kind and class of timber; for unless this is done but a short time need elapse before the same state of affairs that is now confronting our U.S. neighbors will become our own.'[34] From an early date the handloggers were singled out as a particular menace. Roaming the shorelines, they took out only the best sticks of timber before moving on. Moreover, the handlogging system was abused by the sawmill companies, who preserved their timber land held under lease for future use, while purchasing logs from handloggers operating on the public domain to feed their mills.[35]

The dissenting critique was a minority voice in British Columbia, but its influence was not negligible. Since the time of Governor James Douglas, the government was concerned about timberland falling into the hands of speculators. Long-term timber leases were granted only to companies that had sawmills, and the size of the timber lease was linked to the capacity of the sawmill. Critics argued that too much timber was being turned over to a few individuals, but anti-speculation principles guided government timber-allocation procedures. The creation of short-term timber licences, which were open to smaller operators that did not own sawmills, and the licensing system imposed on handloggers in 1884 reflected the ideas of the dissenters.

So, too, did the stumpage laws passed by the legislature, and Suter gave grudging support for the 1884 measure: 'This is not the first attempt to make the saw-mills pay for the timber they stole from the people, but hitherto without success. The bill will no doubt be rendered of little or no use for the purpose desired, but if the bill is passed at all, it will be the thin edge of the wedge introduced.'[36] The more radical proposals, to curb the growth of large sawmill companies with extensive timber holdings, while encouraging opportunities for small-scale capitalists, and to stop timber exploitation because the lumber companies could not afford to pay the government a decent return for Crown timber, were ignored.

The dissenters' critique was also evident in politics. In the colonial period, lumbermen such as John Muir, Captain Edward Stamp, J.A.R. Homer, and Hugh Nelson served in the legislative councils and assemblies. With Confederation the lumber interests continued to engage in politics at the provincial and federal level. J.K. Suter, however, felt that the presence of the lumber industry in politics was unhealthy. He was especially critical of the two large export mills on Burrard Inlet in the early 1870s, and the 1874 federal election prompted a heated debate. J.A.R. Homer, a candidate who was closely connected to the Moodyville sawmill, was labelled a stooge of the monopolistic lumber interests by Suter. Suter's caustic attacks on Homer and the lumber companies were too much for Hugh Nelson, an owner of the Moodyville sawmill: he physically assaulted Suter after the election and was found guilty in a court case.[37] Homer was defeated by a vote of 162 to 142, and Suter was gleeful that the 'timber barons' had lost: 'they absolutely boasted that they would send whom they pleased to represent us in Parliament. Such impudent assumption, was rather too much for our citizens; it was the last straw that broke the back of their patience, and they made up their minds that neither lumber mills nor whiskey mills would dominate in this district ... we are firmly convinced that the vaunted influence of the lumber mill at the Inlet, has melted away like the mist, never to be seen again.'[38]

However, Suter's pronouncement was premature. In the 1875 provincial election, Robert Dickinson, the New Westminster candidate who, according to Suter, was the nominee of the government, the favourite of Moody & Company, and representative 'of the clique that aspires to rule this city,' was victorious. Ebeneezer Brown, associated with the Moodyville operation, was also an election winner. Suter's attacks on perceived monopolists in politics continued through the 1880s.

The dissenters' ideology shaped British Columbia political culture in the nineteenth century. In New Westminster, John Hendry had political ambitions, but as his business interests became more extensive, his political future dimmed. He was elected as city councillor in 1879. In 1886 he failed in a re-election bid because he was a partner in the New Westminster & Port Moody Railway Company, which was asking the City of New Westminster to take over its charter for $800. The conflict of interest was too brazen. In 1889, Hendry was the president of the board of trade and the mayor of the city. Shortly after taking over as mayor, he was forced to resign because of his interest in the Westminster Southern Railway Company, which had dealings with the city. Hendry continued to be politically active, becoming president of the local Liberal–Conservative Association in 1895 and using his extensive connections to lobby governments for favours to benefit his companies, but his business interests made him unsuitable to the electorate as an elected official.[39] Anti-monopoly sentiment had resonance in the province's political culture, even in an era noted for its resource giveaways and government sympathies for big business.[40]

Over the course of the 1890s the new conservation ideas began to circulate in the province.[41] Dr Hugh Watt, member of the legislature representing Cariboo, addressed the provincial legislature in 1894, calling for a select committee to study forestry in the province, and he justified his proposal with a lengthy presentation that articulated the new language and knowledge base of the continental conservation movement. He began by discounting the belief that British Columbia forests were 'practically inexhaustible,' calling for the conservation of the present forests and the replanting of denuded areas. The forests were not the property of the present generation, or of any generation, but rather a 'sacred trust' given us to use, but also to transmit to future generations: 'We are but temporary trustees of the land and its resources.'

Drawing on literature from Ontario and Germany, he outlined the disappearance of the forests in Europe and India, noting that deforested areas affected the climate, rainfall, and the flow of rivers. Agriculture was undermined, but even mining suffered as it was too costly to import timber for construction into reforested areas. While Europeans had made mistakes in the past, they had learned from their mistakes, and Germany, especially, provided a model for forest management in British Columbia. He noted the creation of special schools in Germany and France to train foresters, and was at pains to point out that proper forest management was profitable economically for governments. He spoke of initiatives in the United States in the 1890s that had led to the setting aside of forest reserves to be managed by the government, and he was adamant that private ownership of timber was inappropriate: 'under private ownership the state gets nothing more than the first cost for the sale of the lands and the private owner having only his own personal interests in view is reckless of the evils he may entail on posterity by the too rapid conversion of the standing timber into golden eagles. It is impossible to get private owners of land or of forest, excepting in the rarest instances, to look forward beyond their present interests or wants, or perhaps those of their children.' He also noted the connection between healthy forests and the survival of fish and game, lamenting that in a few years without proper forest management British Columbia would no longer be known as a sportsman's paradise.[42]

Watt was calling for a scientific approach to forest management, where the government managed the forests, monitored logging practices, and regulated the cut of timber, allowing the growth of young trees to match the amount of mature timber logged. The government would be more actively involved in the industry, using specially trained experts to implement proper forest procedures. Unlike the critics of previous decades, Watt's vision was seemingly oblivious to debates in political economy: questions of monopoly and democracy, the role of local communities and individuals, the distribution of wealth and political power, and a vision of a good society were seemingly irrelevant to the technocratic perspective.

The depression of the 1890s allowed British Columbians to postpone grappling with the limitations of forest resource policy, but rapidly changing market circumstances after 1898 led to big forest companies and the government increasingly adopting the language and techniques of the conservation movement. The critique of the forest industry articulated by the dissenters of earlier decades, rooted in anti-monopoly and belief in the search for the ideal political and economic community, did not totally disappear, but with the emergence of the scientific conservation movement it was even further marginalized.

6

Independent Logging Companies/ Timber-Tenure Troubles

The first decade of the century was a dynamic, expanding period in the history of the provincial forest industry. There were more mills and logging camps, and production increased dramatically. There was also a rush to acquire provincial standing timber. Lumbering became the province's most important economic sector. Events beyond the borders of British Columbia contributed to this activity. The disappearance of forests in central Canada and the U.S. Midwest, and the success of the U.S. conservation movement in reserving forested areas, caused investors to turn west in search of standing timber. Trees became more valuable on the west coast. In the U.S. Pacific Northwest, the value of timber was evident when the Weyerhaeuser interests purchased 900,000 acres of magnificent timber from the Northern Pacific Railroad in 1900 at $6 per acre, one of the largest land transfers in U.S. annals, 'a transaction that startled the West and heralded a new era in the lumber industry of North America.'[1] Coastal British Columbia felt this rising interest in standing timber. Despite the tariff, it was believed that timber was going to increase in value in the long run. Then there was the generally buoyant lumber market after 1898, especially in the prairie West, which also influenced the coastal logging and timber business. To meet the demand the rate of logging increased, as did the number of companies participating in the industry. The rapid rise in the numbers of logs being cut, new government regulations, and the transfer of vast amounts of timber to private interests raised concerns and caused consternation among participants in the industry.

One approach to understanding the political economy of this period is to focus on the creation of the Forest Act in 1912. This measure consolidated forestry legislation in one act; created a new Forest Branch, still part of the Department of Lands and Works, to replace the office of the Timber Inspector; and promised adherence to the principles of conservation. W.R. Ross, the minister responsible for forestry in the province, waxed eloquent on the government's achievement:

> An epoch, sir, is drawing to a close – the epoch of reckless devastation of the natural resources with which we, the people of this fair young Province, have been endowed by Providence – those magnificent resources of which the members of this Government and this Assembly are but the temporary trustees. That rugged, rudimentary phase of pioneer activity is doomed to an end. The writing is on the wall; the writing – to put the simple fact – is in this Forest Bill. Armed with that weapon, forged by this honourable Assembly, the Government of British Columbia will undertake the work of forest conservation ...[2]

Historian Thomas R. Roach argues that, because British Columbia was late in embracing the ideas of the conservation movement, it was able take advantage of experiences in other parts of Canada and the United States to put together very progressive measures. Examining the thought and behaviour of politicians and forestry experts, as well as outlining the history of the industry, he concludes that good ideas and good men converged on British Columbia in the years before 1912, and that 'the history of forestry in the province was in many ways a capstone to the early twentieth century conservation movement.'[3] In the Roach argument, the good sense of professional foresters prevailed. Faced with public pressure and feuding sectors in the forest economy, the provincial government 'was forced into the role of arbitrator between the various forest industry factions,'[4] calling a Royal Commission on Forestry. This commission, which met in 1909 and 1910, gave conservationists a window of opportunity, and a sympathetic government was able to impose progressive forestry legislation in 1912.

Martin Robin, less concerned with the ostensible achievements of the conservation movement, focuses on the behaviour of the government in the early years of the century. He argues that Premier Richard McBride's Conservatives, in a grab to get revenue to aid capitalists in developmental projects, changed the timber-tenure system in 1905. With the new system, government revenue increased dramatically, but timber barons rushed to grab control of the province's forests. Robin argues that the bulk of forest land fell into the hands of large corporations before the government was forced to end this privatization of the people's timber in 1907.[5]

This study is less convinced than Roach of the persuasiveness of conservation ideology and of the significance of the new Forest Act in instituting changes. Moreover, unlike Robin, here the government goal of revenue generation is not seen as primary in shaping forest policies. The provincial government remained committed to industrialization and the ideas of the scientific conservationists were used, when appropriate, by participants in the lumber industry to further their specific goals. But business interests, rather than government or scientific agendas, drove change.

The place of the government and conservation is best understood by examining the rise of the independent loggers, logging-camp operators who controlled timber but did not own sawmills. These independent loggers – to be distinguished from contract loggers, handloggers, and mill loggers – emerged as significant players after 1900, with their own needs and demands. The government's policy against log exports galvanized the independent operators, but they were unable to change this government commitment. In timber-tenure policy they were able to move the government, and while government revenue requirements and conservation ideas were factors here, responding to the needs of the independent loggers specifically, and the logging business more generally, was paramount.

Examining the independents also alerts us to a structural change in the industry. The general trend in the development of industrial capitalism in this era was towards greater integration of production facilities and the consolidation of capital in larger corporate entities. In the coastal lumber industry, however, there was increased, although hardly complete, separation between milling and logging operations.

In the nineteenth century there were three types of logging operations, three ways in which mills arranged to get timber for their facilities. First, sawmill companies ran their own logging camps. They owned the timber rights, the livestock, and the machinery; supplied provisions from the company store; and hired foremen to oversee the operations. For example, Robert McIntosh was the foreman at the Dominion Sawmill Company's camp at Mud Bay in 1884. In 1884, the Moodyville Sawmill Company operated six logging camps, and Hastings ran five, in various parts of the coastal region.[6] Second, mills secured a limited number of logs from handloggers, groups of two or three workers who felled Crown timber along the beaches and collected them in booms to be transported to the mills. Handloggers were tied to the sawmills. Mills provisioned handlogger outfits and guaranteed to take their logs. Without such an arrangement, handloggers would be subject to the whims of a fluctuating market when their logs were ready for sale to the few mills that existed. In later years, when handloggers had to have government licences, sawmills often paid the licence fee.[7] Handloggers, then, were not independent entrepreneurs, but rather employees roaming the coastal shores in search of small pockets of good, accessible, unclaimed Crown timber.

Third, sawmills entered into arrangements with contract loggers to take out timber for the mill. In this scenario, the logging contractor owned the livestock and machinery, and was responsible for hiring and managing the crew. In December 1884, the Dominion Sawmill Company entered into an arrangement with the contract logger James E. Murne from Semiahmoo Bay: 'We agree to take and Messr. J.E. Murne agrees to furnish four hundred thousand feet of [illegible] merchantable logs boomed up ready for towing in Nickleomytl [*sic*] River on or about the 20th February 1885 at the rate of four dollars & eighty seven ½ cents per [thousand board feet] ... Dominion Govt stumpage to be paid by us.'[8] In most instances the contract loggers were hired to take out timber owned or leased by the sawmills. Contract loggers ran full-scale operations, the equivalent of the big camps run by the mills themselves, and had extensive capital investments.

Some contract loggers earned substantial wealth. Jeremiah Rogers was the first prominent contract logger. A native of St. Andrews, New Brunswick, he arrived on the west coast in 1858, and, after a short stint on Puget Sound, began work for Captain Edward Stamp, first at Alberni, then at Port Neville, and finally at Burrard Inlet. In the 1870s he continued to run his own logging camps and acquired his own steamer to transport log booms. He died in 1879 at the age of sixty-one, and according to newspaper reports he died wealthy.[9] Angus Fraser, a native of Bay Chaleur, Dalhousie, New Brunswick, arrived in British Columbia in the early 1870s. In 1882 he operated a logging camp on Burrard Inlet and two logging camps on the North Arm of the Fraser River. He purchased a farm on Sea Island for $5,500 in 1881, owned the Richmond Canning Packing Company, and was part owner of the steamer *Ada*. In 1882 he employed between seventy and eighty men in just his farming and logging operations.[10]

In the 1880s the provincial government introduced a timber licensing system, which allowed loggers to control their own timber, but, as we have seen, there was no rush on the part of logging operators to take up timber. Without a competitive market for logs, which would come only when there were sufficient competing milling facilities willing to buy logs, holding timber made little economic sense. It was better to operate on a contract basis. But after 1898, due to changing market circumstances, independent loggers increasingly began to acquire government timber through Special Timber Licences.[11] In coastal British Columbia only 22 Special Timber Licences were taken out in 1898, but this number grew to 248 by 1902 and continued to rise.[12]

According to one estimate, some 407 million feet of logs were cut in the coastal region during 1907, and of this total 289 million, or 71 per cent, was put in the water by independent operators and contract loggers.[13] While it is unclear how much was logged by independents on timber that they controlled, it was sufficient to create a log market in the coastal region. Logging operators without manufacturing facilities sold their logs to mills that would pay the best price, and mills, some of which did not do their own logging, purchased logs at the lowest price possible. The Supervisor of Scalers, Andrew Haslam, said in 1909 that the mills bought considerably more than 50 per cent of their logs from outside sources.[14] In 1906 and 1907, during a log shortage, the Alberta Lumber Company wrote letters to camps up the coast, scrambling to secure logs from independent operations.[15] Mills with their own logging operations also sold to other mills: in 1912 Thurston–Flavelle of Port Moody did not have an established log-buying organization and procured logs from Hastings Saw Mill camps on Thurlow Island and at Rock Bay, as well as from the independent logging company Brooks, Scanlon & O'Brien.[16] Vancouver log brokers facilitated the flow of logs between camps and mills.

The market for logs and the market for lumber did not always move in tandem. With independent operators producing for the log market, overproduction could drive down prices and force camp closures, while the mills profited from low log prices and steady demand for lumber. As the Timber Inspector noted, in July 1911 sawmills were prospering while logging operations were in the doldrums: 'This was caused by an oversupply of logs and not by the dullness of the Lumber Market as some people appeared to think. As a matter of fact all the Mills on the Coast have been running full time and a great number of them running time and a quarter, which is 12½ hours per day. The Millmen report that the market is good and in healthier state than at any time during the year.'[17]

In the log market all participants were not equal. The mills with their own timber and logging camps were much better positioned than companies that merely logged. For mill owners with their own timber and logging camps, the existence of the log market provided flexibility to cope in unstable log and lumber markets. J.S. Emerson, who had his own timber and ran a cedar sawmill and logging camps in 1906, summarized his position as follows: 'I have logging camps of my own and furnish logging camps to the man who has timber, and I furnish timber to the man who has logging camps, and I buy logs occasionally, but not very often.'[18] The independent loggers, though, were much more vulnerable as they had fewer options. They could not play the lumber market against the log market, or choose between logging licenced, leased, or Crown-granted timber. To get leased timber, a sawmill was necessary, and Crown-granted timberland was expensive. With overproduction or dropping log prices, the independents had few options and often faced closure or bankruptcy.

The costs of logging were increasing for independents, contractors, and mill camps alike. Steam donkeys, road construction, timber and labour costs, and the expense of railways were raising the costs of producing logs. T.F. Paterson estimated that it took from $3,000 to $5,000 to equip an ordinary logging camp in the 1890s, but in 1907 it required from $10,000 to $15,000 to do similar work.[19] E.J. Fader opined in 1909 that the cost of logging had increased by $3.00 to $4.00 per thousand board feet over the previous fifteen years.[20] The provincial timber inspector broke down the costs of lumber production in 1911, and estimated that the cost of logging was more than the cost of milling. He estimated the average cost of production at $15.00 per thousand board feet, which included a stumpage payment ($1.50), logging costs ($6.50), towing charges ($1.00), manufacturing and marketing ($5.50), and government dues ($0.50).[21]

The independent logging industry became aware of its special interests at the turn of the century, forming an association to further its interests. The loggers' association's mandate was not to fix prices or establish quotas – this was not feasible, due to the significant presence of the sawmill companies, with their major holdings of private and leased timber, in logging – but rather to lobby the provincial government. Two aspects of provincial policy were especially grating: the log export law and the terms of timber tenure.

The provincial government's policy on log exports pitted the government, often with the support of the mill owners, against the loggers. The government of the United States allowed raw logs to cross the border duty-free in order to encourage manufacturing in the U.S. Pacific Northwest. In the late 1890s more and more logs, especially cedar, were being towed from British Columbia south to U.S. mills. In response, the British Columbia government decided to restrict the export of unmanufactured logs. The provincial government was also influenced by recent popular legislation in Ontario: in 1897, with the support of almost all provincial lumber manufacturers, the Ontario government passed legislation which required that all pine timber cut on Crown lands in the province had to be manufactured into lumber within Canada, adding spruce pulpwood in 1900.[22] Conservation was also an issue as the provincial government wanted to 'forestall a depletion of the timber supply.'[23]

In 1901 the British Columbia Lands Act was amended and a new clause was added: 'All timber cut from Provincial lands must be manufactured within the confines of the Province of British Columbia, otherwise the timber so cut may be seized and forfeited to the Crown and the lease cancelled.'[24] The appeal of this manufacturing condition was clear to both the provincial government and the sawmill and shingle-mill owners. Restricting the export of logs to mills in the United States would force more manufacturing in the province, and thus more capital investment, employees, and economic activity – key goals of the government. For mill owners, restrictions on exports meant that log prices would remain lower because the operators could not take advantage of the U.S. market and supply would remain stable, as the woods would not be stripped to feed U.S. operations. Thomas R. Tait, manager of the Fraser River Lumber Company, made business decisions based on this piece of legislation, which is clear in this 1901 letter to the government: 'We have recently begun the manufacture of shingles here in the City of New Westminster, having decided to build a large shingle mill after the passing of the Act by the B.C. Government prohibiting the export of logs, and would respectfully ask if it is the intention of the Government to enforce the provisions of said Act. If cedar logs and bolts are permitted to be freely exported it will be a hardship to ourselves [as well as to other mills] ...'[25]

The independent loggers rallied to fight this legislation, which limited their market options. In 1901 the provincial logging operators created the British Columbia Lumbermen's Association, changing its name to the British Columbia Loggers' Association in 1904, to clearly distinguish it from the mill owners' organization, the British Columbia Lumber and Shingle Manufacturers' Association. Although mill owners such as J.S. Emerson and E.H. Heaps were at times involved, the B.C. Loggers' Association spoke for the independent loggers. The loggers railed against the new legislation from the outset. They argued that there were not sufficient mills in the province to absorb the output of the logging camps. They struck a conservationist pose, arguing that the British Columbia mills took only the best part of the logs, unlike U.S. mills, and that the provincial government lost 'stumpage on the logs so left in the woods to either rot or add fuel to a fire which may be sweeping over that section.' They also pointed out that it was economically unsound for the government to give support to the milling industry, whose production costs were lower, rather than the logging operations, where more money was expended, and thus more economic activity generated. The provincial government remained committed to the policy, and while they temporarily deferred enforcement of the provision, they also closed loopholes that allowed some logs to be sent to Puget Sound. The original measure was poorly worded, allowing logging operators with Special Timber Licences to export logs, but this was prohibited by an order-in-council in July 1902.[26]

In the summer of 1904, the logging operators launched a full-scale attack on the ban on log exports. The industry was in the doldrums after the boom years of 1902 and 1903, but there were logs sitting in the water that could not be sold to provincial mills. The operators wanted to ship this surplus to the United States. Especially galling were the millions of feet of cedar logs lying in the water at and near Vancouver. Loggers reluctantly accepted the ban on the export of most types and grades of logs, but they wanted to be able to export low-grade cedar logs, which were utilized in the United States but had no market in Canada. When logging operators put in a costly camp, they wanted to take out as much merchantable timber as possible, including low-grade cedar, to cover their costs.

Loggers challenged the export ban in the courts, but to their dismay the government amended the Timber Manufacture Act in 1906 to close the loopholes.[27] The milling interests were pleased:

> The action taken by the Legislature to compel timber cut from our Crown lands to be manufactured at home has been hailed with satisfaction, and the effect has been most beneficial. It may in a way be said to be the turning point in the history of our lumber industry. Previously our lumber companies, handicapped by the competition of Washington millmen manufacturing our logs and sending back the finished product to Canada duty-free, were barely able to make ends meet, and in some instances the local mills were actually losing money.[28]

The export issue, however, did not disappear. The boom year of 1907 generated increased production in the woods, but by early 1908 mills had reduced cutting logs in response to soft market conditions, leaving thousands of board feet in the water. Logging operators were also caught with logs in the water and were forced to cut production. Notably, mill owners as well as logging operators appealed to the provincial government to allow the export of low-grade cedar that had no market in Canada. It was estimated that, if action was not taken, $500,000 worth of logs sitting in the water would be eaten by teredos, the small marine worms that infested and destroyed logs left in the water too long.

The government responded by lifting the embargo on 8 March 1908, allowing the export of low-grade logs. Some 8 to 10 million feet of logs, largely cedar, went to Puget Sound over the next months, leading to increased activity in the woods. But the exemption did not only clear logs from the water, it also led to the reopening of camps, some of which were financed by U.S. mills. The government restricted the export privilege in August, but the suddenness of the action caught many producers unaware, leading to an extension of the privilege until 1 November 1908. Loggers took advantage of the reprieve; in September and October 1908 alone, J.A. Sayward's camp at Powell Lake was able to ship almost a million feet of cedar to the Motor Shingle Company in Anacortes, Washington.[29]

By 1909 coastal millmen with camps and independent logging operators agreed that the export of low-grade cedar, for which there was no market in Canada, would benefit them, but although they continued to press for change, the government resisted. The restriction on log exports was popular with the general public, and the government hoped that the embargo would generate a market in Canada and encourage industrial investment. The government also feared that laxness with regard to cedar would eventually undermine the whole anti-export policy. Premier Richard McBride commented on the policy in the fall of 1910: 'The States appear to be clamoring for raw material. They could have the British Columbia raw material on one condition – that they set up their plants within British Columbia's borders.'[30] Time was also on the premier's side: in early 1911 cedar prices began to rise in coastal British Columbia, even surpassing prices in Washington state, and the shingle mills, which had relied on cedar bolts, were now increasingly taking cedar logs.[31] The ban on the export of raw logs was maintained in the new forestry legislation of 1912.

Table 6.1 Special Timber Licences in British Columbia, 1898–1906

Year	Coast region	Kootenay region	Other interior/ coast north of Vancouver Island	Total
1898	22	41	24	87
1899	33	36	18	87
1900	68	61	14	143
1901	49	71	10	130
1902	248	259	19	526
1903	438	685	184	1,307
1904	350	923	178	1,451
1905	561	1,122	442	2,125
1906	1,494	1,634	818	3,946

Source: Tabulated from Original Licences, BCARS, B 4700, v. 1–3.
Note: Yearly tabulations by the Lands Branch for all of British Columbia differ a little at times from my own calculations. The government figures for each year are in brackets: 1898 (87); 1899 (87); 1900 (138); 1901 (130); 1902 (527); 1903 (1,306); 1904 (1,465); 1905 (2,166); 1906 (3,959). These figures are not broken down into regions: British Columbia, *Sessional Papers* (1898–1907).

The log-export issue paled in comparison with the furore caused by the changes in timber-licence policy. After 1898 coastal logging operators began to take up Special Timber Licences with increasing frequency (see table 6.1).

Although the Kootenays experienced a rush on timber licences even greater than the coastal region, it was the coastal operators who exerted pressure to change the terms of the licences. It was not only independent logging operators with no milling facilities that took up Special Timber Licences; mills that ran their own logging camps also used this form of tenure. For example, in 1899 C.M. Beecher, R.H. Alexander, and R.C. Ferguson of British Columbia Mills Timber & Trading Company; H. McDonald and L.A. Lewis of Brunette Saw Mills; and W.L. Tait, a shingle manufacturer, took out timber licences on behalf of their firms. In 1903 the list of mill owners was even longer, including Tait, Beecher, Alexander, Ferguson, McDonald, Lewis, D.C. Robertson, H. Hackett, E.H. Heaps, A.M. Spicer, A. McNair, John Hendry, E.H. Bucklin, and G. Small.[32] However, since most of the mills also held timber leases, it was the independent coastal logging operators who relied much more on the Special Timber Licences.

The terms of the timber licences attracted criticism in 1901. Licences could be held for only one year, and were renewed at the discretion of the Chief Commissioner of Lands and Works. The insecurity of tenure meant that loggers could not use their timber holdings as collateral when borrowing from the banks. The timber grant could be no larger than 1,000 acres, and the licence was non-transferable, meaning that the holder of the licence could not sell it to a sawmill or logging operator but could only return it to the government. This created hardship: 'a logger might expend $1,000 or $2,000 upon a limit in improving the same, putting roads and ways, etc., but in the event of his becoming insolvent or getting under hard circumstances, he could not sell his limit to another logger.'[33] Finally, no person was entitled to hold more than one licence, and so mill owners and logging operators took out licences under the names of other parties. But this left them vulnerable as these parties could demand payment from the firms or refuse to allow them to log the limits.[34] In 1903 the government made changes, lengthening the term of Special Timber Licences from one to three years, and by then one person could hold two licences. The banks, however, still viewed them sceptically.[35]

The economic slump of 1904 caught many logging operators with too many logs in the water, and the drop in log prices pushed firms to the brink of bankruptcy. To remain solvent and weather the recession, independent logging operators needed the financial support of their bankers. The independent loggers intensified their call to change the terms of the licences. They requested that the term of the licence be extended to twenty-one years and that persons or companies be allowed to hold many licences. They also wanted the licences to be made transferable, meaning that they could be bought and sold instead of merely being returned to the government. In other words, operators wanted the government to create a market for timber licences. This, it was believed, would satisfy the financial institutions. They would loan money to logging operators based on the twenty-one-year licence, and if the logging company failed, the bank would take over the licence and be free to sell it on the open market, thus recouping at least a portion of its loss.

A few independent loggers were wary of the proposals, arguing that this would set the stage for monopoly capitalists to take over large tracts of timberland in the province, but most of the independent logging community supported the plan. The provincial government was amenable in this case, perhaps because they were not willing to offer the independent loggers any relief on the log-export policy. It seems unlikely that the provincial government was aware of the long-term consequences of the proposed changes. Amendments to the Land Act became effective on 15 April 1905. Existing Special Timber Licences remained valid for sixteen years and were transferable, and licences acquired after 15 April would be good for twenty-one years and also transferable.[36] There were no restrictions on how many licences a person or company could hold.

The speculative fever that swept British Columbia in 1906 and 1907 was not solely due to these changes in timber regulations. The new license system was in place by April 1905, but the number of Special Timber Licences issued in 1905 was only 60 per cent higher than in 1904 in the coastal region, and 21 per cent higher in the Kootenays. Increases had been proportionally higher in 1902, as compared with 1901 and between 1902 and 1903. In 1906 the number of Special Timber Licences in the coastal region more than doubled, reaching 3,946 in the province as a whole. Then, in 1907, there was a spectacular rise in the issuance of licences: soon there were more than 15,000 licences registered province-wide, and the number may have exceeded 17,000. As of 10 August 1910, there were roughly 15,000 timber licences in the province; 7,910 were in the coastal region and 7,090 in the Kootenays and other parts of the interior.[37]

The need to rebuild San Francisco after the 1906 earthquake created a great demand for lumber products. There was also a healthy economy in the prairie West, and governments supported new railways across the prairies and through British Columbia. Railway construction projects consumed timbers and lumber, and opened up areas for commercial lumbering. Investors anticipated new sawmills and logging operations in the interior. As one commentator noted, 'Timber would not have been worth a nickel without the Grand Trunk Pacific.'[38] Larger economic trends and a feeling in North America that timber was scarce and destined to become increasingly valuable, coupled with the transferable aspect of the Special Timber Licences, fuelled the demand for timber licences in British Columbia. The rush on timber licences increased government revenue. The government's take from licences rose from $7,150 in 1900 to $177,984 in 1904, to a whopping $1,339,351 in 1907.[39]

It is impossible to break down precisely the purchasers of timber licences. Among the multitude of cruisers and locaters tromping through coastal forests in 1907, two crews in 1907 illustrate the presence of carpet-baggers from Ontario and the close working relationship between 'scientific' professional foresters and private capitalists. Dr Judson F. Clark was an outspoken professional forester who left his position as

Provincial Forester in the Ontario government in 1906. He became associated with a Toronto syndicate that wanted to take up timber and enter the lumber business in British Columbia. Clark became manager of the Continental Timber Company; the president was W.T. White, general manager of the National Trust Company. Prominent Toronto businessman Joseph Flavelle was also involved in the timber syndicate. Clark went to British Columbia and secured twenty-four Special Timber Licences for the syndicate by 1908.[40]

Upon hearing of the plans of the syndicate, John Flavelle approached White and Joseph Flavelle, seeking to join the consortium, but was denied. He decided to set up his own timber firm. Led by Roland D. Craig, a forestry engineer from Ottawa who had been in charge of firefighting in the Dominion Forest Service, this crew included H.R. MacMillan, a young forestry student at Yale, and Aird Flavelle, an eighteen-year-old young man from Lindsay, Ontario. Flavelle would become a sawmill operator in the coastal region in 1912, and MacMillan the Chief Forester of the province, and later British Columbia's most prominent lumber industry capitalist. The party was financed by John D. Flavelle, Aird's father; James Carnegie, a Port Perry lumber merchant; and Robert James McLaughlin, a Lindsay barrister. They put up about $25,000, and Craig had a 10 per cent interest in the syndicate as well as his wages.[41] The Craig and Clark parties worked the coastal area north of Vancouver in 1907, and at times travelled together. There was a sense of urgency to their work, according to Aird Flavelle: 'Dr. Clark says we ought to get hold of everything possible no matter how much this year and possibly next year but after that it will be too late.'[42]

British Columbia timber brokers, coastal mill owners, and independent logging operators also took up Special Timber Licences, as did Americans. There was much comment on the increasing U.S. presence. A large number of Special Timber Licences were sold to Americans, 'vague "American interests," and to readers of advertisements in Chicago and Philadelphia and the east generally.'[43] In British Columbia, still smarting from the 1903 Alaska Boundary Decision, which most British Columbians felt unfairly benefited the Americans, anti-Yankee sentiment remained strong. The idea that Americans, and speculators at that, were gobbling up provincial resources created public concern. Estimates that 75 per cent of the province's timber holdings were under U.S. control have led to claims that 'the influx of capital and entrepreneurs from the United States was the industry's distinguishing feature during the decade after 1900.'[44] The exact proportion of U.S. timber holdings in the province is unknown, and tracking the trading of Special Timber Licences after 1905 is extremely complicated, if not impossible. Licences were traded, but how many were sold by Canadians to Americans and Americans to Canadians is unclear. In this period independent loggers from the United States also became more prominent in the coastal region, but most independent loggers migrated to Canada and took up residence. Julius H. Bloedel, who was to become one of British Columbia's most prominent lumbermen, was established in Washington state, and, anticipating tariff changes, began moving his operations to British Columbia in 1911.[45] But many Americans had Canadian roots, illustrating the difficulty of classifying lumber operators. R.J. Thurston was born in Ontario, but spent fourteen years sawmilling in the Willamette Valley, Oregon, before relocating to British Columbia.[46] Another lumberman, H.H. Spicer, was born in Nova Scotia and worked in the lumber business there before spending ten years in the United States, five of these with a Michigan firm, and then moving on to Vancouver in 1888.[47]

While Americans owned a significant amount of timber, the bigger question is whether this had any appreciable impact on the development of the industry beyond increasing the flow of capital into the coastal region. Did Americans or U.S. practices directly influence timber-tenure policies, manipulate log markets, or impose a particular U.S. stamp on the development of the industry? Generally, it seems that Americans with money showed a propensity for buying timber outright on Crown-grant and Railway Belt land; they did not trust the tenure security of licences and leases. But there was no push or U.S. lobby to extend private ownership of forests. Further, it is not clear if buying timberland was a strategy pursued by Americans or if it was a practice followed by individuals with lots of money, individuals who tended to be American. Nor did American holders of Special Timber Licences pursue a uniquely U.S. line. Indeed, many, like other licence holders, probably fared rather poorly. The Flavelle–Craig timber cruising party, representing Ontario backers who had put up $25,000, staked timber in 1907. The plan was to stake twenty limits, then sell off some in a few years in a rising market, using this money to finance the yearly fees. Then, five or ten years later, they would sell of with a handsome profit. However, it was not until the 1940s that the properties were unloaded. Two reverted to the government during the depression of the 1930s, and eight more were sold off to cover fees that were owed to the government. The price they received for the eight licences recovered only a fraction of what they had paid out in dues to the government over the previous decades.[48] There was no log shortage in this period, and lumber markets from 1905 to 1940 were more often depressed than booming. U.S. investment contributed to the commodification of the forests and to government revenues, but its influence should not be exaggerated.

The rush on Special Timber Licences in 1906, and especially 1907, raised the spectres of speculation, monopoly, and Americanization, and the McBride government was put on the defensive, but within the lumber industry the problem was more technical. By late 1907 there were some 7 million acres in the province held under Special Timber Licences, holding an estimated 90 to 140 billion feet of merchantable timber.[49] Licence holders held the privilege for twenty-one years, and thus to recoup their investment in rental fees the holders would have to log off their land within twenty-one years. It was not guaranteed that their licence would be renewed. All this timber, then, would be sent to the mills within twenty-one years, and likely would be logged most aggressively in years when log prices were high. This raised serious problems. The milling side was not capable of handling this volume. Moreover, the forced deadline would generate the overproduction of logs as licence holders rushed to liquidate their holdings, forcing prices down, destroying forests at less than their market value, and undermining the long-term viability of the industry. The future promised chaos for the industry and the provincial economy, as well as an inefficient, uneconomic harvest of British Columbia forests.

Another concern of those holding Special Timber Licences, especially those actively engaged in logging, was that, despite the twenty-one-year term, the banks still did not look on the licences favourably. As William Murray, manager of a Vancouver branch of the Canadian Bank of Commerce, noted in August 1909: 'as a rule we do not consider a timber license a tangible security.'[50]

With future problems becoming increasingly clear, licence holders held a meeting in Vancouver in the fall of 1907, leading to the formation of the British Columbia Timber and Forestry Chamber of Commerce on 27 November 1907. This organization has been characterized as merely the voice of speculators, those who purchased timber licences in expectation of turning them over at a profit in the future.[51] More accurately, the association represented all licence holders who recognized the long-term implications of the timber-licence form of tenure, bringing together independent logging operators and mill owners, as well as people holding timber on speculation, to solve a rapidly worsening problem as more and more licences were being issued.

At the inaugural meeting, John Hendry was made honorary president, and the executive included a number of people active in logging and sawmilling, such as E.H. Heaps, M.N. Garland, D.J. O'Brien, Joseph Chew, T.F. Paterson, J.S. Emerson, Captain Theodore Magneson, George Davenport, and L.L. Dickerman. Distinguishing speculators from businessmen is not easy; many logging operators sold licences if the right price was offered, and some timber holders decided to enter the logging business. However, the association represented the interests of licence holders and they came from a variety of positions in the business community. Dr Judson F. Clark, representative of the Continental Timber Company and trained forester, was elected secretary-treasurer and became its spokesperson.[52]

The provincial government's first response to the unfolding timber-licence problem came in late December 1907, when it placed all unleased and unlicensed Crown land under reserve. No more Special Timber Licences would be issued. The problem remained as to what to do about existing licences. The British Columbia Timber and Forestry Chamber of Commerce offered its solution. Most importantly, they wanted the terms of Special Timber Licences extended beyond twenty-one years to allow for an orderly harvesting of the timber crop. Licence holders should, they argued, have the rights to an area covered by a timber licence until the timber on it was logged off.[53] For the government, giving perpetual ownership to present holders of Special Timber Licences was problematic because the public would perceive it, correctly, as a windfall for the lumber interests and speculators already holding licences.

By 1907 there were a number of other problems in the coastal forest industry that had been exacerbated by the boom. Purchases of large sections of private timberland by wealthy individuals and companies led to a perception that a few people would soon control the resource and push up the price of logs and lumber. That many of these publicized deals originated in the United States also fuelled concern. This land had been previously alienated by the provincial government, much of it in the Esquimalt & Nanaimo Railway Belt, and it was beyond provincial taxation, royalty, and export laws. Independent loggers were concerned about speculation and monopoly: 'we feel that it will not be in the best interest of our province that the development of our immense timber resources should be concentrated in the hands of a few large companies able to restrict and control at their discretion this development so as best to advance their own personal ends.'[54] If possible, logging and mill operators cushioned themselves from the threat of rising log prices and speculated by acquiring large tracts of timber themselves from private sources. In this period a few large timber sales constituted 100,000 acres, while most were less than 50,000 acres.[55] At the same time, many sawmill companies held timber leases, which were cheaper and almost as secure, since they were easily renewed. British Columbia Mills Timber & Trading, for example, held 92,590 acres under lease in 1910, North Pacific Lumber Company held 69,597 acres, and Brunette Saw Mills held 15,583.[56] Standing timber had become extremely valuable and participants in the industry tried to secure as much timber as possible. The Thurston–Flavelle cedar sawmill operation acquired as much standing timber as they could, but they did not log this timber. Rather, they bought logs from independent loggers and hoarded their own timber for the future: 'This situation was to remain the operating custom on the coast for many years.'[57]

Large operations tried to secure timber from a variety of sources, including privately owned land or Crown grants, provincial leases, dominion leases, and Special Timber Licences. In 1909, A.D. McRae's massive Fraser River Lumber Company held 30,000 acres of Crown-grant timberland, 5,000 acres of Dominion timber land, and 40 Special Timber Licences in the coastal region. The independent logging company Brooks, Scanlon & O'Brien held 8,000 acres of Crown-grant land, 30,000 acres of Dominion timber land, 25,000 acres of provincial leases, and 40,000 acres under Special Timber Licences. The Rat Portage Lumber Company of Vancouver held 11,000 acres of Crown grant in the Esquimalt & Nanaimo Railway Belt, which they had purchased in 1906 for $10 an acre, 30,000 acres under Dominion lease, and 80,000 acres under provincial lease. The Ladysmith Lumber Company on Vancouver Island held 5,000 acres under Special Timber Licence and about 1,100 acres of Crown-grant timberland. Small & Bucklin, a New Westminster sawmill and logging company, held 2,000 acres under Dominion lease and some 15,000 to 20,000 acres under provincial lease. Some companies, of course, relied on just one form of timber tenure. The Canadian Shingle Mills Company logged 12 Special Timber Licences; Nimpkish Lake Logging worked 43 provincial timber licences on northern Vancouver Island; and the International Timber Company logged 60 square miles of Crown grant, purchased from the E&N Railway Belt for $10 an acre, near Campbell River.[58]

The government keenness to encourage the development of the pulp-and-paper industry in the province also raised concerns. Pulp-and-paper projects demanded heavy capital investment, and generated jobs and capital economic activity. Moreover, pulp-and-paper mills used wood fibre, not logs, and thus used more trees, and a higher portion of individual trees. This fit with the government's goal of more efficiently and economically harvesting timber and with the conservation ethic, which recommended taking out all old-growth timber and raising a new crop of trees under the guidance of foresters following scientific principles. The pulp-and-paper entrepreneurs, for their part, wanted guaranteed access to large quantities of wood fibre before investing in costly facilities. In 1901 the provincial government introduced twenty-one-year pulpwood leases. Independent loggers and mill owners were alarmed as the pulpwood leases included logs suitable for sawmills. The issuing of leases was discontinued in 1903, but by this time four leases, comprising 354,399 acres, had been granted, and a further 1,300,480 acres had been placed under reserve for future use by the pulp-and-paper industry. The first wood-fibre pulp mill was not constructed in coastal British Columbia until 1909, but it was clear that the new industry was increasing the pressure on the resource.[59]

Handloggers were another problem. Since the 1880s handloggers received a licence from the government to take out logs from any land that was not privately owned or held by lease or licence. Handlogger licences were cheap, a mere $10 per year. In the first years of the twentieth century, two aspects of the handlogging system alarmed many. First, handloggers took the best, most accessible, timber from an area. This was unsound from the perspective of the conservation movement, and if in future a mill company or logging company wanted to take up a licence on the same area, the value of the tract was diminished because high-quality timber had already been logged. Second, handloggers were no longer groups of two or three men taking out logs using primitive tools. Some loggers with handlogger licences were running full logging camps, using steam donkeys to yard logs from deep in the forest. Lease and licence holders claimed that this was unfair as the handloggers paid little for the privilege compared with their financial outlay for licences and leases. The handlogging system was also abused by the sawmill companies: the Rat Portage Lumber Company, for example, used handloggers to take out logs specifically for them, making the handloggers essentially employees of the sawmill firm.[60]

In 1906 the provincial government imposed technological limitations to restrict the operations of handloggers: 'The holder of the license under this section shall not use horse, cattle or steam power in carrying on lumbering operations under such license.'[61] The legislation did not stop the handloggers; 408 handlogger licences were issued in 1906, increasing to 573 in 1907. The government introduced more draconian legislation in 1908, restricting handlogging to the area north of Rivers Inlet up the mainland coast, at a higher fee of $25 per year. Rivers Inlet, beyond the northern tip of Vancouver Island, was outside the economic and geographic unit of the coastal lumber region because it was too far away, and log booms faced the perils of open ocean. Handloggers no longer contributed substantially to the mills in Vancouver. In 1908 only thirty handlogger licences were issued.[62]

A range of other issues faced the government and industry. Farm groups complained that logging practices denuded land along creeks and rivers, leading to floods; worried about the impact of logging on the water supply; and charged that the extension of timber leases and licences slowed agricultural settlement.[63] Destruction of forests by fire was a concern shared by all, but the issue of who was responsible and how to control it was not clear: hunters, prospectors, settlers clearing land, sparks from railway locomotives and donkey engines, campers, and lightning were all deemed culprits in varying degrees.[64] Logging practices were also criticized. From the perspective of conservationists, too much merchantable timber was being left in the woods as companies driven by the market took only what would fetch the best economic return; it was estimated that at least 25 per cent of a tree was left in the woods.[65] Logging with donkey engines was criticized for starting forest fires and for destroying too many young trees, thus stemming the regeneration of the forests. Andrew Haslam, Supervisor of Scalers and former logging and mill operator, stated the issue bluntly in 1909: 'Steam logging is a much greater menace to the forest than team logging, on account of the danger from fire and the greater number of small trees that are torn down; at the same time there are many tracts of good timber that could not successfully be logged with teams that are profitably logged with steam.'[66]

The immediate problem facing the government and the industry in 1907 and 1908, though, was what to do about the Special Timber Licences. Extending the terms of licences beyond twenty-one years was a foregone conclusion, but it would be politically unpopular, except to stakeholders in the industry, and would bring to the fore other grievances against the industry and the government's forest policy. The solution was the calling of a royal commission in 1909 to investigate the industry and make recommendations for changes. A three-man commission, headed by Frederick J. Fulton, the provincial Minister of Lands, held hearings throughout the province, attended mainly by lumbermen. The government pressured the commission to deal quickly with the Special Timber Licences, and an interim report was produced recommending that Special Timber Licences be held as long as there was still timber on the land. This recommendation was translated into legislation before the final report of the commission was placed before the legislature in January 1911. The Fulton Report incorporated many of the ideas and principles of the 'scientific' conservation movement, although this conservationist thrust was moderated in the government's legislative response.

Independent loggers, mill owners, timber speculators, and the public were content with the government's forest policy of 1912. Indeed, lumbermen were at the forefront of conservation. In January 1911, John Hendry, president of British Columbia Mills Timber & Trading, returned from attending the meetings of the Canadian Forestry Association and the Dominion Conservation Commission, leading national conservation organizations, in central Canada. On his return he met with Premier McBride to discuss timber and conservation matters.[67] The details of this conversation are not known.

It was clear that any belief that forest policy would be guided by scientific principles expounded by trained, dispassionate experts was puffery. Markets and the interests of business prevailed, as they had in previous years. Policy decisions remained with the cabinet, whose decisions were shaped by the financial interests of loggers and millmen, and the dynamics of the market.[68] The depression of 1913 to 1915, the economic dislocations of the First World War, and the post-war recession of the early 1920s ensured that the conservation ethic was put on the back burner. In the late 1920s and the 1930s, the weaknesses of the 1912 Forest Act and the powers of the Forest Branch launched a new wave of public criticism.[69]

After 1898 more and more logging companies were independent of milling firms. A buoyant lumber economy increased demand for logs, and there were sufficient milling operations to create a log market. The independent loggers were especially vulnerable to market fluctuations, bank demands, and unsatisfactory timber tenures. The anti-export policy on logs was generally supported by millmen but not by the independent loggers. It limited their options and squeezed them financially. The independent loggers battled against this legislation to no avail. However, they received government help with new terms for timber licences, leading to an unexpected crisis in the timber industry after 1905. The solution was yet another change in the terms of the timber licences. The Forest Act of 1912, which amalgamated policies of previous years and wrapped them in the rhetoric of conservation, satisfied interests in the industry, reflecting the ongoing policy of the state, but the main issue of timber tenure had already been solved.[70]

The rise of independent logging companies emphasizes the fact that coastal lumbering was still very competitive in 1913, embracing a range of entrepreneurial firms. The push to integrate milling, timber-holding, logging, and pulp-mill operations into large corporate entities did not begin in earnest until the 1930s. Forestry giant MacMillan Bloedel was the epitome of the age, bringing together two already large integrated firms, Bloedel, Stewart & Welch and the H.R. MacMillan Export Company, into one corporate entity in 1951.[71] The age of dominance by a few integrated firms in the forestry sector had finally arrived.

7

Industrialization

Popular histories of the forest industry in British Columbia have been fascinated by the machinery, techniques, and workers in coastal logging. Books, full of pictures, show logging railways, steam logging machines, and grapple yarders, using illustrations to demonstrate ground-lead logging and high-lead logging, as well as more modern techniques.[1] This interest reflects the pride of participants in the industry, as well as a fascination with machinery and organizational techniques. The focus on the coastal region, where the trees are larger and the equipment thus different from that used in other parts of North America, also suggests pride of place, a celebration of the West Coast as a unique environment, where men were forced to overcome major obstacles to produce socially useful goods.

The popular logging histories are valuable records, but their discussions of equipment and technique are devoid of broader contexts. The relations between men and machines, the roles of capital and labour, and the imperatives of profit-making are largely unexplored in understanding the creation and sustaining of the production process. As such the similarity of logging with the larger process of industrialization is obscured.[2]

While logging has attracted popular and scholarly attention, the milling side in the production process has been neglected. As we have seen, the same capitalists operated both milling and logging operations, and the absence of studies of milling is odd. The mills, too, faced special circumstances in coping with large logs and had to develop new techniques and equipment, distinguishing the West Coast from other parts of North America. Perhaps the outdoor environment of logging and the human interaction with nature appeals to a particular masculine vision of freedom, while the noisy, dirty mill environment allows for few positive symbols. Possibly, and more controversially, the prominence of Asian workers in the British Columbia mills by the turn of the century discouraged the dominant, non-Asian society, the creators and controllers of popular culture until recently, from appropriating this work landscape and incorporating it into their version of history.

This chapter looks at machinery, techniques, and the organization of production processes in both milling and logging. The focus is on industrialization. Recent scholarship emphasizes the complexities and diversity of industrialization in different economic sectors and in different regions. Still, as Craig Heron points out, the development of manufacturing suggests some basic characteristics:

> The form could vary between industrial sectors, but in manufacturing the long-term transition was from household or artisanal workshop to factory. Implicit in the term are new technology, power sources, and management practices, larger units of production, and in many cases a concentration in urban areas ... In particular, the industrial revolution is credited with accelerating the formation of new social classes, especially the working class, and creating the conditions for serious social conflict.[3]

Coastal mills were employing steam power by the 1870s, and the woods operations began using steam machinery extensively a decade later. Moreover, both operations had elaborate processes involving different machines and many human beings, processes that had to be coordinated and managed. In the new large sawmills of the early twentieth century, electric power and automation were prominent features. This chapter looks at the machinery and techniques of the production process. The next two chapters discuss the workers. In all three chapters the pace and extent of change was influenced by the fluctuating demands for products in the marketplace.

In early operations, logging and milling were closely linked spatially, taking place in close proximity to each other. In the 1860s the Stamp operation on the Alberni Canal and the various mills on Burrard Inlet all took out logs that were near the sawmills. In small operations logging and milling production systems continued to be intertwined, drawing on the same labour force. Reginald Pidcock engaged in lumbering in the Comox Valley in the 1870s and 1880s. Catering to the local market, and mixing farming and grinding grain with sawmilling, Pidcock operated the mill until 1885, when he sold out to meet debts.[4]

Pidcock's diary in the early 1880s documents the relatively simple, fragmented production process of the small operator. On 8 March 1882, he sawed all day, but had constant trouble with the belts in the water-powered mill. His wife, Alice, who was helping him, 'got her foot a good deal hurt with the log carriage.' On 30 March 1882 the big belt in the mill broke and he had great trouble getting the pulley off. The next day no sawing was done: 'getting off pulley all day, had to break it up at last.' It was very cold on 11 April, but he sawed steadily even though the logs were bad. He spent 13 April grinding saw teeth in the morning and sawing lath in the afternoon. On 25 April his nine-year-old son, Willie, and his seven-year-old son, Harry, helped him in the mill. On 1 May, the 'cows got over the long bridge all morning finding them. Sawed in the afternoon.' In January and February 1883, Pidcock and hired help logged. On 16 February 1883, he 'went to work with the oxen to haul logs. Had to mend the road in a few places, but did very well the ground being very hard and the logs hauled better without the sleigh than with.' He spent 14 March 1883, grinding grain, and on 4 May he 'ran the planer for a while in the morning & ground grain in the afternoon. Paid Tom off who does nothing but grumble. Little Mary [his six-year-old daughter] had a narrow escape from lumber falling on her.' On 6 July 1883 'Bob the Indian came to work & we sawed till noon,' and five days later, 'Willie got his leg rather badly bruised with the carriage in the mill running over him.' On 12 July the tides were very low and 'in the evening Alice helped me haul a log up by starting the mill ... Willie's leg better.' On 12 September he went down to the steamer after dinner and 'got up two Chinamen, one a boy for the house & one to work in the mill.' On 19 October he dismissed the Chinese houseworker, 'who refused to work,' and by 24 October the Chinese millworker had quit. On 10 November 1883, he 'sawed in the morning. In the afternoon settled up with the loggers after a deal of trouble. They [are] leaving for good this afternoon.'[5]

In larger operations, milling and logging became increasingly separated geographically as the industry grew. As the timber near the mills was felled, and the demand for logs increased, logging operations began to extend up the mainland coast and through the Fraser Valley. In the late 1860s pairs of handloggers were working on Howe Sound, and in 1873 there were '5 or 6 small logging camps, where men were working, the only signs of life about this celebrated Sound.'[6] Handloggers were also active on Pitt Lake in the early 1870s; steamers towed the logs down the Pitt and Fraser Rivers, and then on the ocean to the mills on

Burrard Inlet.[7] By 1875 the logging frontier had extended up the coast to Jervis Inlet, where a Moodyville Sawmill camp and First Nations people, the Sechelt, took out logs.[8] The northward march continued, and in 1884 Moodyville Sawmills had a camp on Texada Island.[9] The logging perimeter also expanded to the south and the west. Directly to the south of New Westminster, a series of camps was established on Mud Bay and Semiahmoo Bay in the early 1880s, using the Nicomekl River and Campbell Creek as waterways. To the west logs were coming from Stave Lake and Stave River.[10] Camps were also established on the North Arm of the Fraser River, at False Creek and Port Moody, and on Valdes, Salt Spring, and Gambier Islands.[11] The sawmills at Burrard Inlet, New Westminster, and Port Moody were on the water, and thus were connected to all the logging camps to the north, south, and west. Logs from Howe Sound went to New Westminster and logs from Pitt Lake went to Burrard Inlet. In 1882, W.P. Sayward took a boom of logs from New Westminster to his sawmill in Victoria.[12] However, the New Westminster mills tended to draw from Pitt and Stave Lakes, the north Arm of the Fraser, and Mud and Semiahmoo Bays, while the Burrard Inlet mills took logs from the North Arm of the Fraser, English Bay, Burrard Inlet, and the northern mainland coast.

In the years from 1886 to 1892, the number of logging camps in the coastal region increased, and the frontier of logging activity was extended. In 1889 there were some twenty large camps, each with between twenty-five and forty workers, up the mainland coast and on the east side of Vancouver Island. The Hastings Saw Mill camp and store nine miles above the Yuclataw Rapids was the most important centre in the northern area, 'it being the point of supply for all the company's camps, timber cruisers, and also for the numerous hand loggers working in the area.' Hastings had three camps in the area. There were also camps to the north of the Hastings establishment, stretching as far as Port Neville. One logger estimated that there were more than 700 loggers working in the region north of the Fraser River.[13] Up the Fraser River to the east, large camps operated by George Blackstock and the Brunette Saw Mill Company were established on the Lillooet River, which flowed into Harrison Lake, by the first years of the 1890s. Coastal-region camps employed between 1500 and 1700 men in the late 1880s.[14] The depression of the 1890s curtailed the geographic advance of the logger. After 1898 the establishment of increasing numbers of camps resumed.

Logging involved two processes. First, the trees were felled, using saws and axes, then cleared of limbs and cut, or bucked, into log lengths. Second, the logs, or sticks, were transported to a body of water. This hauling, or yarding, was at the heart of coastal logging, and consumed the most time, energy, and cost. Coastal logs were massive. The Hastings Saw Mill camp at English Bay took out a log in 1883 that measured 40 inches in diameter at the top, tapered gradually to the butt, and was 112 feet long. A Pitt Lake logging camp took out a log in 1885 that was 110 feet long and had a 52-inch diameter at the top, or small end.[15] Handloggers circumvented the difficulties of manoeuvring heavy logs by falling the trees directly into the water. Primitive tools were sometimes used by handloggers to move logs a short distance to the water: George Ward DeBeck claimed to be the first 'to put a log in the water using a jack screw.'[16]

To log timber that stretched back from the water or the mill site, more extensive procedures were involved. Paths and skid roads were constructed into the forest, and the logs were dragged out, or skidded. Preparing skid roads took planning and cost money. In 1882 the Keeler camp at Crystal Creek, on the Pitt River just before reaching Pitt Lake, 'was compelled to remove a ridge of rocks by blasting ... to construct a substantial skid road.'[17] Such an expenditure was worthwhile only because there was sufficient timber to ensure the road's use for five years. Yokes of oxen supplied the power to yard logs in the pre-1885 era. Jeremiah Rogers imported oxen into the Alberni Valley in 1860 to work in logging, and his short-lived spar camp at Port Neville in 1864 and 1865 employed eight work oxen.[18] Even with oxen and skid roads yarding was difficult and costly. The limitations of animal power, physical impediments, and the rough terrain made the yarding of each log an adventure.

At an early date, too, devices such as chutes and flumes were used to move logs. Here is description of a dry slide in operation at Moodyville in the first years of the 1870s: 'The timber ... reaches Messrs. Moodie [*sic*] and Co's mills at Burrard Inlet by means of an immense dry slide half-a-mile in length, which gives passage to the enormous saw-logs ... Leaving behind them a long train of smoke, they plunge into the deep water, causing the spray to fly up into the air some thirty feet ...'[19]

Logging-camp operators constantly searched to improve yarding, and both steam power and rail lines were experimented with on main skid roads in the pre-1885 period. In 1874 Jeremiah Rogers used a road steamer at his English Bay camp to replace oxen in pulling logs. According to one commentator, the steam-powered vehicle 'commenced a new era in logging in British Columbia ... [since] it worked like a charm, and hauled out saw logs with ease that ten yoke of oxen could not have started.'[20] He was unduly optimistic as the road steamer was in the end no replacement for oxen. After Rogers died in 1879, his operation was purchased by James Gillies, and Gillies continued to search for more effective ways of logging. In 1882 he installed a primitive logging railway at the camp:

> There is a good, substantial wooden railway, about two miles in length, extending back into the forest, and upon this is successfully operated one of Thompson's (Glasgow) traction engines. By the ingenuity of Mr. Mooney this engine has been adapted to the work by removing the rubber bands and chain armory from the driving wheels and substituting two double flange wheels in front. Mr. Mooney has also constructed a very strong car for carrying the logs, many of which are of great size. It is provided with eight wheels, without cross axles, so that the wheels, acting on a pivot, readily adjust themselves to the track and seem to answer admirably. The car takes about 20,000 feet of lumber at a load, and makes from 4 to 5 trips a day.[21]

In 1885 Angus Fraser experimented with steam power in one of his logging operations up the mainland coast by having a large scow fitted with powerful steam engines: 'The idea is, when a log is cut down, prepared and rolled into a suitable position, a wire rope will be attached, connected with the engine in the scow, and by this means the log will be hauled to the water, thus saving road-making, cattle, etc.'[22] Another prominent logging contractor, James Gillespie, put in a tramway at his logging camp at False Creek in 1883. The wooden track was 8-foot gauge, and the rails were 12 inches square. Mule teams were used to draw the cars loaded with logs, replacing the five yoke of oxen previously employed at the camp. Obviously such an operation would work only on flat terrain. The tramway extended two miles into the woods from the water at False Creek. Gillespie operated two other logging camps on the North Arm of the Fraser River; one used four mule teams and six yoke of oxen, and the other employed fourteen yoke of oxen.[23] Logging operators were well aware that improved yarding procedures could lower their costs of production, and the operators of substantial enterprises invested time and money to upgrade their hauling systems. Yet while steam and rails were both used in woods operations, oxen remained the primary source of power, and it was not until after 1885 that full-fledged railway logging, with steel rails and locomotives, was introduced.

The economic boom after 1886, coupled with the large volume of timber per acre in coastal forests, justified building costly logging railways. In 1888, Royal City Planing Mills constructed a logging railway at its Mud Bay camp, the steel rails coming by ship around Cape Horn. The locomotive had been used in the construction of the Canadian Pacific Railway (CPR). In 1890 the company also had a logging railway at English Bay. The steam locomotive used in the operation was manufactured in Kingston, Ontario. In 1890 Royal City Planing Mills was running three steam locomotives. Other large firms followed suit. Maclaren and Ross built a seven-mile-long logging railway in 1892 from the Fraser into its timber east of Sumas Mountain, and in 1894 the Moodyville Sawmill Company began constructing a logging railway into its Grief Point timber.[24]

By 1908 about fifteen logging companies operated logging railways in the coastal region, and they had about 106 miles of track laid in the woods. British Columbia Mills Timber & Trading alone had five railway camps: the New Westminster operation had 7.75 miles of track, the Thurlow Island camp had 3 miles, the Rock Bay camp had 13 miles, the Salmon River camp had 5.5 miles, and the Granite Bay camp on Quadra Island had 5 miles of track. The Victoria Lumber & Manufacturing Company had 21 miles of track around Chemainus, and other railway shows included those operated by the Brunette Saw Mill Co., Anderson Logging, and Heaps Timber.[25] In 1911, according to one estimate, there were thirty-one logging locomotives in British Columbia, overwhelmingly on the coast, and a tour up the mainland coast and the east coast of Vancouver Island by the Timber Inspector a year later noted a few of the camps which had railways pushing from five to fifteen miles into the woods, including the Brooks, Scanlon & O'Brien camps at Stillwater and Kingcome Inlet, the Stewart & Bloedel camp at Grief Point, the Canadian Puget Sound camp at Powell River, and the International Timber camp at Campbell River.[26]

Well-capitalized firms with substantial timber holdings built logging railways deep into their limits, but the railway was a fixed transportation link, and logs still had to be hauled to an accessible railway siding. Steam donkey engines, which were cheaper and more portable, were a solution for short hauls. Moreover, they were eventually used by operators of smaller enterprises who could not afford a logging railway, and a series of donkey engines could be strung together to get logs out from deep in the woods. The principle of the steam logger was simple: a cable was run into the woods, and one end of the cable was attached to a log and the other end wrapped around a spool. A steam engine drove the spool, thus drawing in the cable and the attached log. Refinements in the machine took place over time.[27]

The first effective use of steam to yard logs in British Columbia was likely at J.M. Stewart's Secret Cove camp in late 1891. Drawing on an idea pioneered in San Francisco, Stewart had British Columbia Iron Works of Vancouver construct a floating logging camp. A large scow housed loggers and was home to a steam engine that powered the scow and hauled logs to the shoreline.[28] In December 1891 Stewart reported that 'the steam substitute for oxen ... is a success in every particular.'[29]

Many in the lumber industry were sceptical of the merits of steam power and referred to Stewart's Folly, freely predicting that it would be a failure. In the late spring of 1892, Stewart's machine passed a test that convinced doubters:

> It is no uncommon thing for a steamer in a storm to lose a boom. In the past all efforts to get a boom that had gone ashore into the water had proved futile and thousands of dollars have been lost this way every year. A month or so ago a steamer was caught by a morning gale just off Point Grief, and soon it became a case of let go the boom or be dragged ashore with it, and the boom was allowed to go. In the morning the captain hunted out Mr. Stewart and asked him if he thought his new fangled machine could get the booms off again, and being offered a fair recompense went up to the scene. The boom was made up of very large logs and some of them lay in a very poor position for handling, but the machine hauled them all into the water again in short order. One old logger who had predicted certain failure and whose faith had been pinned to oxen watched the machine's way of walking the big logs around in utter amazement. One huge piece of timber he was sure could not be hauled over the rocks and stones that lay between where it had been left by the tide and water, but when the chain was attached to it it slid into the water without a hitch. 'That settles it,' said the old logger, 'all the oxen I could have got into line could not have hauled that log over those stones.'[30]

Soon boilers and engines were being mounted on sleds and dragged into the woods. The Victoria Lumber & Manufacturing Company is usually credited with using the first steam donkey in the woods, sometime in 1892, and by November 1893 there were four steam logging operations in British Columbia, and it was anticipated that they would soon become ubiquitous: 'These machines to an extent revolutionized the trade, and it is expected that before long they will have entirely supplanted oxen in the logging camps of British Columbia.'[31]

There were a number of reasons why the steam donkeys appealed to logging operators. They were more powerful than oxen and brought out logs that were previously inaccessible: 'With his steel cable and windlass and powerful engine, the steam logger can yank the huge giants over ground when oxen could not be used at all.'[32] The donkey engines needed fewer workers and often there was no need to build and maintain skid roads: 'Four men with a steam logger can do as much as 10 with oxen, and in the majority of cases with the steam logger there is no need at all to build a skidway.'[33] The donkey engines were also faster than oxen and they were less costly to maintain. In 1896 the Brunette Saw Mill manager 'sold the firm's aging logging cattle because they had been "eating their heads off" and went on to prescribe the purchase of steam donkeys "as the cattle is too slow to log to any advantage." Moreover, he pointed out "the fuel is on the ground [i.e., wood] ... and when we shut down for the rainy season all the expense could be shut off."'[34]

The steam donkey did not suddenly replace animal power in the woods. The depression of the 1890s drove down the cost of labour, curtailed expenditures on equipment, and halted the expansion of logging operations. In camps that continued to operate, steam power and animal power often coexisted in the same operation. In 1895 Walter Gilley ran a twenty-eight-man camp about three miles from New Westminster, using both steam donkeys and animals in yarding. After trees were felled, cut into proper lengths, and stripped of branches, they were pulled by a steam donkey to a skid road. The skid road was one and a half miles long, and about $5 worth of grease and oil were applied to the road every day to prepare it for the logs. The logs were dragged along the skid road by ten-horse teams, managed by a teamster, the highest-paid worker in the camp: 'The horses are guided by word of mouth only, and appear to understand every word said to them. For instance, in starting a heavy haul, and wishing to get the animals away with a steady pull from the word "go," the driver will say, "Straighten out a little," and the next instant each horse will stretch forward until the traces are as tight as fiddle strings. Another word, and the horses are off, dragging at a fast walk sixteen or eighteen tons of logs over the rough skid road.' At the end of the skidway the logs were dropped into a 600-foot-long chute, down which they slid into a deep ditch. The logs were then floated down this mile-and-a-half ditch to the North Arm of the Fraser River, where they were boomed, ready to be towed to the sawmill.[35]

Donkey engines got larger. In 1903 Vancouver Engineering Works built the largest logging machine in the province for British Columbia Mills Timber & Trading. The engine weighed 17 tons, had 10-inch cylinders with a 15-inch stroke, and the strong frame was built with 15-inch steel I beams.[36] In 1911 there were some 353 donkey engines being used in the woods of the province, mainly on the coast.[37]

Steam power was also used in new ways. In 1911 ground logging, whereby by a series of donkey engines hauled logs along the ground, began to be replaced by high-lead logging, where a central steam engine was linked by cables through two high spars to one end of a log in the woods, and this end was lifted and the log dragged to a central cache. Developed in the United States, high-lead logging did not necessitate the construction of skid roads, as the logs were yarded from a wide circumference around the central spar tree. While the system was more efficient and faster in hauling out merchantable logs, it destroyed young trees and the forest floor in the affected area. With the trees swinging through the air, it was also more dangerous for workers. The Canadian Western Lumber Company camp at Courtenay is credited with first employing the technique in coastal British Columbia.[38]

As logging costs rose, and timber became scarcer and more valuable, operators increasingly tried to squeeze out as much profit as possible from woods operations. Closer attention to inventory and cost accounting helped. Logging companies began employing more sophisticated managerial techniques, and accountants and office managers took on a greater role in directing the production process in the woods. Some of the older logging operators were slow to respond to the changing value of timber. R.H. Alexander was associated with Hastings Saw Mills since the 1870s, and the company held much standing timber. In 1912 Alexander was flabbergasted when Bloedel, Stewart & Welch purchased timber for a high price from the government that he believed Hastings had logged out some years earlier. When he looked into it, Alexander found that there was an error in the company's records and only half the limit had been logged. Thereafter, Hastings recruised every one of their old properties.[39] Companies became much clearer as to their timber holdings, cruising their timber limits to ascertain the exact volume, the different tree species, and the terrain. Extant records from the Canadian Puget Sound Lumber Company in 1912 show a precise record of the company's extensive timber holdings, broken down by type of tenure and giving the board feet of each species of merchantable standing timber on the different limits.[40]

Arrangements between timber holders and logging contractors were also becoming increasingly precise. A 1910 agreement between Anderson Logging and the Victoria Lumber & Manufacturing Company set the terms by which Anderson would log Victoria Lumber & Manufacturing timber on Vancouver Island. The agreement was nine pages long, stipulating all the obligations of the contractor in legal language, as well as the price the mill would pay for the logs to be harvested.[41]

In 1913 the Victoria Lumber & Manufacturing Company revamped its accounting procedures. An outside accountant was brought in to examine the company's operation, make recommendations, and then comment on the changes that the company made: 'Your system of records of woods operations is greatly improved over previous years ... Camp payrolls showing proper distribution of departmental operating costs being kept at each camp, is going to facilitate greatly in keeping in closer touch with the actual costs of logging.'[42]

In 1907 the Nimpkish Lake Logging Company was preparing to purchase the Edmonton Logging Company, which was taking out logs at Nimpkish Lake on northern Vancouver Island. The potential purchasers knew the value of the assets of the Edmonton Logging Company and the specifics of their timber holdings. But they sent a Victoria chartered accountant, W. Curtis Sampson, to Nimpkish Lake to calculate productions costs. Sampson spent one day watching the fallers and tabulating their performance, and another day watching the yarding crew, which was using one donkey engine. The crew of two fallers felled eleven trees on 15 March 1907. The trees contained from 1,148 to 7,044 board feet, and the total day's production was 46,705 board feet. The yarding crew, which consisted of eleven men, took out 46 logs, ranging from 16 to 64 feet in length and containing a total of 74,752 board feet, on 16 March 1907. Sampson recognized that the workers did the very best that they could because they were being scrutinized, but his experience in studying other logging operations suggested that they would comfortably produce 40,000 feet per day on average. Sampson detailed the wages of each worker, including the foreman, cooks, and blacksmiths; allowed for depreciation on the equipment, fuel costs, and the price of accident insurance; and calculated that the cost of daily production was $174.80 to produce a 40,000-foot boom of logs at salt water. He also calculated the operating costs for a proposed logging railway. In the end Sampson predicted that logging would yield a net profit of $3 per thousand feet. Nimpkish Lake Logging, which included British and local investors, was pleased with the analysis and purchased the operation.[43]

What is remarkable about the data is the knowledge that owners and management possessed about the production process in the woods. Armed with this information, management in Vancouver or even London could set production quotas which workers would be expected to meet. Low production would cause investigation by company officials. Detailed cost-accounting procedures and the techniques of scientific-management movement associated with Frederick Winslow Taylor, whereby managers appropriated the knowledge of the workers at the worksite to rigorously control production, were present in the coastal forest industry.[44] Logging-camp foremen and crews faced not only the physical problems of falling and moving timber in difficult physical environments, and the impositions of new machinery and technical logging systems, but also the quantified imperatives set by management, based on management's awareness of the nature of work in the woods. Management techniques, as well as machinery, were reshaping the production system in the woods. Although evidence is sparse, there were suggestions that production in the woods was becoming faster and more efficient. As a Vancouver correspondent noted in 1908: 'The logging camps, although fewer in number and with smaller gangs of men, are turning out a larger proportionate production than they did a year ago.'[45]

Once logs had been put in the water they were collected into booms and towed to their destination sawmill. Sawmills, logging contractors, and independent captains owned and operated the steam tugs, which took provisions to the camps and returned with logs. For the most part, the journeys were uneventful, but at times, due to rough weather or operators' mistakes, booms were lost or broken apart and tugs were damaged. In 1869 the *Isabel*, owned by the British Columbia and Vancouver Island Spar, Lumber, and Sawmill Company, ran aground just beyond the First Narrows in Burrard Inlet, causing extensive damage. In 1883 a scow loaded with provisions from the Moodyville Sawmill Company for one their camps was lost in a storm, at a cost of about $1,500.[46]

Once the logs were delivered to the sawmill, they were manipulated through a series of processes to transform them into lumber of various dimensions. In the 1860s and 1870s the coastal sawmills became predominantly steam-powered. The Muir sawmill at Sooke had installed steam power in the 1850s, securing the machinery from a wrecked steamship in 1855, and the Stamp mill at Alberni was also steam-powered.[47] But the early mills at Chemainus and on Burrard Inlet were water-powered, and in the mid-1860s mill owners such as W.P. Sayward and S.P. Moody were petitioning the government for permission to build dams and redirect creeks to ensure a steady flow of water during the dry season.[48] The Hastings mill, which began operation in June 1867, was steam-powered, and in the same year Moody converted to steam power. In future years steam was the norm in all coastal mills. The first DeBeck Bros. sawmill was powered by two 12-horsepower steam engines, but in 1884 the improved mill was driven by one 40-horsepower engine and one 75-horsepower engine, using three boilers. The Moodyville sawmill was run by a 500-horsepower steam engine, using eleven boilers, and the Royal City Planing Mills operation used two 200-horsepower engines.[49] A newspaper description of the Moodyville operation in 1876 gives a sense of how the power was distributed through the sawmill: 'The mill itself is a mammoth building upwards of 300 feet in length, 270 of which is roofed over and covered for the greater part with corrugated iron. The interior represents a complete network of shafts, bands, and wheels, by which the machinery is kept in motion, the whole thing being driven by a powerful steam engine, which formerly belonged to H.M.S. *Sparrowhawk*.'[50]

In the mills the logs passed through a series of saws. A journal observation captured the Moodyville sawmill in action 1875:

> A log hauled up out of the boom into the mill, is first rolled by means of hooks & chains & friction gearing overhead, onto a travelling cradle. There arranged in position & wedged up, & then run through the saws. Outer slab taken off & rejected, then two or three planks, Log then turned on the flattened Side & again run through & perhaps turned yet again after a few boards have been taken off, So as to manage to get the greatest possible quantity of clean lumber from the outside. Log advanced after each cut by a pair of screws worked each by a man, distance regulated by the character of log & sort of lumber required.[51]

The first cut was made by circular saws in larger mills in the 1870s. The big saw at the first DeBeck Bros. mill cut 28 inches, and 'in order to cut big logs would run the saw through the center, turn log back and cut one quarter, turn back again and so on until the log was used up.'[52] In 1884 the mill used two 54-inch circular saws, which was the average in large sawmills.[53] The circular saws squared the log by cutting off slabs, and further cuts produced planks of smaller dimensions. In the 1870s and 1880s, some big mills used gang saws to reduce squared timber to boards: the Moodyville sawmill in 1882 had 'a gang saw, carrying twenty-five blades, going all the time, making twenty-six boards at a time.'[54] Edgers and trimmers were saws that cut planks to the appropriate length and width.

Once the lumber was cut to the desired dimensions, there were a number of options. Rough lumber was either stacked in the yard; sent through the planer to produce smooth, dressed lumber; and/or taken to kilns, where it was heated to take out the moisture, which lessened its weight and prevented warping. Finally, in the export mills, lumber was loaded into ships.

The inclusion of sash-and-door and furniture manufactories in large mill operations added another complex stage in the production process after the late 1870s. The manufactories consisted of a wide variety of power-driven machines in one building. In 1883 the manufactory at Royal City Planing Mills housed eight circular saws, band and jig saws, four planing machines of various kinds, moulding machines, shaping machines, a morticing machine, a tenon machine, a sand-papering machine, a panel-raising machine, a shingle machine, and a turning lathe.[55] A description of the manufactory at Dominion Sawmills, which produced sashes and doors as well as furniture in 1884, drives home the point that woodworking, like sawmilling, was becoming increasingly mechanized in a factory system:

> In the filing room the sparks emitted by the steel when it comes in contact with the emery stope, looks like a shower of stars. A twenty five horse power machine moves the innumerable wheels when they are in belts. The planer, a very plain-looking machine, is a great worker. Into it a steady hand introduces a board; it moves slowly to a hand on the other side, and when it is through it is planed, and as smooth as ivory. The endless bed moulder is fit to make in a minute twenty-five copies of any moulding; and beside it stands the Railway wood clipper, that is used for cutting wood into short pieces of any shape. The panel planer is a perfectly finished thing, and the band saw a wonder – a wee little thing that walks through a piece of wood and makes, if you will it, a thousand turns in a minute. And there is the shaping machine ready to suit itself to your ideas,

> and turn wood into any shape, if your hand moves according to the order which the mind should give. The lathes, mortice makers, and the sash and moulding machines are great curiosities to those who never saw such things, but the machine that smooths the wood with sand-paper is a gem. The cramp-presses are powerful machines.[56]

After 1886 mill owners introduced larger and more sophisticated machinery into their facilities. In 1890 Royal City Planing Mills built a new drying house which used a system that dried lumber faster: an air chamber was heated by coils of steam pipes and from this the hot blast was sent through the building by means of a fan worked by steam. In 1894 Brunette Sawmills imported a 13.5-ton planer from New York that was capable of planing a piece of lumber 16 × 30 inches on all four sides on one pass through the machine. The major change, though, in sawmill technology in the early 1890s was the introduction of band saws to replace circular saws to make the initial cuts in logs. The advantage of the band saw was that it had a lesser kerf than the circular saw, thus preserving wood that normally was turned into sawdust. I.M. Kendall, who devised many new devices for the coastal logging and sawmilling industry, designed the massive band mill that was installed at the Maclaren–Ross operation in 1890. The mill weighed 35 tons, and the band saw itself was ten feet in diameter and the blades were ten inches wide. Kendall also installed a band mill at Cassady's on False Creek, and with a kerf of less than 1/8 inch it saved 15 per cent in lumber compared with the old circular saw, which had a kerf of 7/16 inch. Brunette Saw Mills installed a band saw in 1894 to secure the same advantage: 'This saw only cuts out an eighth of an inch in sawdust, instead of three-eighths by the circular saw, thus saving one inch board out of every four cuts.'[57]

In the first decade of the twentieth century the push to create a continuous line of production in the sawmills intensified. Despite the ongoing transformation of the production process, the changes came 'so gradual that their importance upon the great lumber industry of the country is apt to be overlooked.'[58] Mill owners increasingly employed steam power to move logs, lumber, and waste through the facilities to speed up the process. A system of steam-powered rollers, transfers, and conveyors removed the need for manually handling the product. In the sawmill constructed by E.H. Heaps & Company in 1907, the lumber throughout the mill was 'handled entirely by automatic steam transfers, live rollers and other labour-saving devices.' There was also a powered conveyor system for handling all waste material.[59] Similarly Fraser River Sawmills updated their large sawmill in 1906 and much attention was given to facilitating the flow of the production process:

> The saw mill is equipped with a complete system of conveyors, trimmer-saws, transfers and live rolls. It is also equipped with a large air compressor, which furnishes the power for operating the power setworks on the carriages on both sides of the mill; also the kickers for rolling logs off the jack ladder, for lifting all transfer skids, and for raising the edger rolls ... This is probably the first saw mill to adopt this system, and those in charge consider it a great success, as well as a convenience.

Before entering the mill, logs were cut into the required lengths by a steam drag-saw which was set up on a float in the river, a novel innovation. The steam for the drag-saw was conveyed through a pipe and a hose from the boiler house. When the lumber left the mill, special care was taken to ensure that the lumber followed a downward course until loaded onto railway cars or ships.[60]

The sawmill at Harrison River, owned by the Rat Portage Lumber Company, epitomized the modern mill in 1910:

> The mill is so thoroughly equipped that from the time the log leaves the water to enter the mill not a hand is placed upon it. Huge grab hooks, ponderous chains, and massive centres under the control of a man at levers do the work in a few seconds. Instead of the laborious work entailed in tail sawing, in which all the sawn lumber is handled and piled by hand, endless chain carriers from every saw carry the lumber to the edgers and sizers, and from there to the lumber yards. There is a continuous stream of lumber, and the capacity of the mill is about 125,000 feet per day.[61]

Mills also were moving from steam to electricity as a source power. Previously electricity had just been used for lighting. Moodyville Sawmills had introduced electric lighting in February 1882 by putting up ten lamps at a cost of over $4,000. Night work was now possible, and the hours of production increased.[62] But in the first years of the century, electricity was also used to power machinery. By 1907 about a dozen smaller sawmills in and around Vancouver were powered by electricity. In 1910 the massive sawmill operated by the Canadian Western Lumber

Company was using electricity to drive many of its machines, and in 1911 the new J. Hanbury & Company mill on False Creek was entirely driven by electricity.[63]

Changes to the production process in the woods and in the mills were ongoing in the years from 1858 to 1913. Steam power quickly replaced water power in the large milling operations by the 1880s. In the woods, steam power became prominent after 1885, but the larger operations that relied on logging railways and steam donkeys coexisted for a period with logging operations that continued to rely on animal power. The rate and timing of the introduction of changes to the production system were related to broader market considerations. The economic booms in the years from 1885 to 1892, and again after 1900, spawned the introduction of new technologies, while the depression years of the 1890s did not. The increasing market value of timber after 1900 led to closer management of logging and milling to ensure full value was taken from the resource.

Logging and milling were not primitive staple industries using simple techniques, even in the nineteenth century. As Richard Rajala concludes, 'during the late nineteenth and early twentieth centuries, west coast logging underwent its industrial revolution.'[64] The mills were even more advanced by the early twentieth century, perfecting the application of electricity and smooth-flowing production processes that were increasingly automated. By the time finished lumber left the mill, the wood had passed through a series of processes, beginning with the falling of the trees in the forests. At each stage, techniques were applied to the product, transforming the economic worth of the product and adding value, and these individual applications were coordinated and linked in an elaborate system. In sash-and-door plants, even further refinements took place. Like textile plants, shoe factories, agricultural implement manufactories, and steel mills, coastal logging shows and milling facilities were a dynamic, active part of industrializing Canada.

8

Loggers

The British Columbia coastal forest industry is notable for its lack of union and political activism before 1913. Vancouver Island coal miners, Kootenay hard-rock miners, and skilled urban workers in Vancouver organized unions and mobilized politically, but loggers and millworkers were not prominent.[1] Millworkers engaged in some strike action from the 1880s onward and attempted to build unions, but no permanent structures were established. Loggers were even less militant, with little evidence of union drives. This is curious. Loggers and millworkers were industrial labourers working in factory-like settings, constantly subject to the push of owners to increase productivity, hostage to volatile markets and corresponding wage fluctuations, and responding to new machines and changes to the production process. Work in the woods or the mills was no less strenuous or demanding than other working-class occupations. In the woods, men were isolated in primitive conditions and subject to the whims of the camp bosses. Surely, this state of affairs was conducive to militancy and radicalism.

The reason for the lack of organization is partially explained by linking race and skill as features of the workforce that undermined solidarity. Robert A.J. McDonald, in his social study of Vancouver, concludes that 'skill and race intersected in ways that presented a formidable barrier to class mobilization among lumber workers ... Consequently, in Vancouver's most important industry the working class was sharply divided against itself.'[2] In the mills, skilled white workers held the best jobs, and were supported by a lesser cast of poorly paid, unskilled workers. After the 1880s Asian workers constituted this secondary group, and thus race and skill intertwined. After initially trying to exclude Asian workers, the white workers actually participated in the racialization of the labour force, benefiting from the presence of a poorly paid, racially defined pool of unskilled workers.

In the woods, race was not an issue: loggers were largely white. However, a job hierarchy that rewarded the more crucial workers in the production system and earned a degree of employee loyalty, and the geographical dispersion of the workforce in isolated camps, undermined solidarity. Further, the lack of a union tradition in other lumbering regions that coastal loggers could draw on, the frequent downturns of the log market, and a logging culture that extolled the independence of loggers made organization difficult.

This chapter and the next discuss workers in the production system. Job hierarchies, ethnicity, race, age, and marital status, as well as unions and strikes, are discussed. Moreover, issues such as getting paid, the dangers of the work, perceptions of loggers, and the construction of a notion of masculinity that included union participation are also examined. This chapter looks at loggers, and the next deals with millworkers.

Logging camps varied in size, but in the 1880s, the pre-steam era, the average size of a large camp run by a sawmill or logging contractor was about twenty-five men. This included a foremen or boss, a cook, a blacksmith, a number of men engaged in falling and bucking timber, and the yarding crew. The most important and best-paid workers in the camps were the teamsters. They were responsible for maintaining and working the oxen, as well as supervising the yarding. They made the important decisions about where and how a log was going to be dragged from the woods. In 1864 teamsters were paid $100 per month, while axemen received $70 to $80. By 1882 the discrepancy between the wages of teamsters and axemen had increased: at Burrard Inlet camps teamsters received $125 per month, and axemen $60 per month.[3]

The steam age introduced new categories of workers into the woods. Steam engineers, hook tenders, and firemen were necessary at all large operations, and teamsters over time gradually disappeared. The wage schedule at Victoria Lumber & Manufacturing camps, which were railway operations, illustrates the job hierarchy in the woods. The locomotive engineer was paid up to $75 monthly, and the rest of the workers were paid by the day. Hooktenders were paid up to $4.00 per day, while section foremen, undercutters, head fallers, head loaders, blacksmiths, and sawfilers received up to $3.50 per day. Workers receiving between $2.75 and $3.25 per day included donkey engineers, head skidders, second fallers, buckers, second loaders, conductors, and roadmen.

Locomotive firemen, rigging slingers, brakemen, snipers, head barkers, line horsemen, boom men, and head swampers were paid between $2.50 and $2.75. The lowest rates, between $2.25 and $2.50, went to blacksmith helpers, spool tenders, second barkers, second swampers, and signal men.[4]

If an aim of employers in moving to steam power was to undermine the wages and influence of teamsters, they were not successful. Steam donkeys in yarding made teamsters, the elite of the woods workforce, redundant, but created a new class of skilled labour – the donkey engineer. As F.H. Parks of International Timber noted in 1911:

> The use of donkey engines in the woods is, of course, familiar and is similar to all Pacific Coast operations. The attitude of the Government in its regulations as regards boiler inspection, both of locomotive and donkey engines, is severe. Frequent inspections are made and licenses are required for all classes of engineers. This has a direct and far-reaching influence on the labor question and results in increasing trouble and expense creating as it does another class of skilled labor to be reckoned with.[5]

Worker response to the changes in the production process were not noticeably hostile. The transition to steam in the woods took place over an extended period of time, and in the early twentieth century oxen and horses were still used in some camps. Teamsters, then, could still find work.

Wages fluctuated with the availability of labour and the price of logs, and the two were interrelated. In a buoyant economy, workers had opportunities to work at many jobs, and log prices were high. Wages in these circumstances went up. However, pay cuts came with economic downturns. The slack market in early 1908 after the record year of 1907 forced wages down:

> Fallers can now be secured at from $2.75 to $3.25 a day, while they formerly got $4.50 and $5, and were scarce at that. Hook tenders now work at $3.50 and $4 a day, and board themselves. Formerly they demanded $5 and $6 and board. Engineers formerly got $90 a month and board, and they now work for $3.50 a day, and board themselves. With the loosening up of money and the revival of trade, there is no doubt that wages will have a tendency upwards ... When the demand is brisk lumber must be cut at any price, and the 3,800 loggers in British Columbia are not numerous

> enough to go around; when the demand has fallen off there are hundreds of loggers practically begging for work at any price. So it is the demand for lumber regulates the wages of loggers.[6]

Loggers were well aware of log prices, which were public knowledge, and adapted to the changing circumstances, enduring hard times and personally negotiating for higher wages in stronger markets.

The manuscript census for New Westminster District offers insights into the workforce in the coastal lumber industry. The New Westminster Census District included the city of New Westminster, the north and south sides of the Fraser River, Burrard Inlet, and the mainland coast north of Burrard Inlet. In other words, it was the coastal region excluding Vancouver Island and islands in the inside passage between the mainland and Vancouver Island. The manuscript census is not perfect, as some census takers referred to all Chinese and Native workers as labourers, without providing specifics. Other occupations also cause difficulties. Skilled workers were often just designated as millwrights or blacksmiths, without adding whether they worked in sawmills, logging camps, or non-forest-related facilities. Still, despite these limitations, a reading of the manuscript census gives a perspective on the workforce that is not available elsewhere.

The 1881 census identifies 165 loggers in the New Westminster District, and the 1891 census 786.[7] The median age of the loggers was 34 in 1881 but had dropped to 30 in 1891. In terms of place of birth, the logging workforce was predominantly British and Canadian in the nineteenth century: 76 per cent had been born in these two places in 1881, and about 80 per cent in 1891. Ontario was the main Canadian source of loggers in both census years. Only a small proportion of non–First Nations loggers were married, 4 per cent in 1881 and 7.9 per cent in 1891. Nineteenth-century loggers were predominantly of British and Canadian origins, young, white, and single.

While the median age of First Nations loggers was roughly the same as for non-Native loggers – thirty in 1881 and thirty-four in 1891 – their marriage rates were not similar. In 1881 there were twenty-three First Nations loggers identified in the census, Sechelts at Jervis Inlet, and all were married. In 1891, of the sixty-eight First Nations loggers, 84 per cent were married.

Census data likely underrepresent the presence of First Nations and mixed-blood peoples in the logging workforce. According to one observer, in the mid-1870s Native handloggers on Jervis Inlet put over a million feet of logs into the water in one year for delivery to the Moodyville sawmills, and another group of Native workers further to the north exceeded that level of production. They were paid $3 per thousand board feet.[8] Roughly calculated, this contribution may have accounted for 30 per cent of the this mill's log supply. In the 1870s Native people on Jervis Inlet wanted their own timber lease, but had to be satisfied with a commitment from the Moodyville sawmill that a portion of the mill's lease would be set aside for Native people to log.[9] Native people worked independently of the company-run camps, and as handloggers were less directly controlled by the discipline of company foremen and company-devised production systems.[10]

Of 19 general timber licences for handloggers issued in 1886, 14 of the recipients were identified as Native people; in 1887, 11 of 17 were Native; and in 1889, 19 of 55 were Native.[11] In the summer of 1894, Indian Agent Devlin visited every reserve up the coast, noting the increasing logging done by Native peoples:

> The majority of the Indians cultivate potatoes and other vegetables, and of these they have plenty to last them through the winter. Instead of spending the fall and winter months in idleness or hunting, as was their habit a few years ago, the Indians are now turning their attention to handlogging, and they have several camps already in operation. The logs are readily purchased by the saw mills, and in this way the Indians are able to earn enough money to supply them not only with necessaries, but luxuries.[12]

First Nations peoples protested the encroachment of logging operations in their home areas, and chafed at paying licence and stumpage fees. They also demanded that timberlands be set aside for them to log. Whites, for their part, blamed Native peoples for many forest fires, arguing that they were set 'for the purpose of driving the loggers away.'[13]

Woods crews remained overwhelmingly white in the twentieth century,[14] although many camps used Chinese cooks. The Victoria Lumber & Manufacturing Company used Japanese men in grading roads in the woods, and employed Chinese workers as water carriers, supplying the steam donkeys.[15] But the general consensus among logging-camp operators was that Asians were not suited for work in the woods. As R.H. Alexander of Hastings Saw Mill said in 1901: 'Neither Chinese nor Japanese are used in the woods. They are not suited for it.'[16] The government was well aware of the politics of anti-Asian sentiment in British Columbia. In 1906 the provincial government added provisions against the employment of Japanese and Chinese workers to the Special Timber Licences. The provision was not well enforced until 1911, when protesters called on the government to clamp down on the employment of Chinese and Japanese workers on government timberland. In 1912 the government launch *Kiora* was 'patrolling the Gulf and inspecting the camps at Howe Sound, Jervis Inlet and as far north as Desolation Sound, ... making inspections of the Japanese camps in order to secure evidence to be used by the Deputy Minister for the cancellation of the Licences upon which those Japs are working.'[17]

Chinese and Japanese workers were rare in camps taking out logs, but they were active in other areas in the woods. Japanese crews were dominant in taking out timber for making oars, and in producing shingle bolts for manufacture in shingle mills. One Japanese shingle-bolt camp in 1909 had two Japanese operators on Howe Sound, taking out bolts for the Chew Lumber and Shingle Company. They built roads, a dam, and a 700-foot flume to take out bolts from high on a mountainside that was viewed by many as impossible to log.[18] Chinese workers had long predominated in producing cordwood. A railway traveller in 1887 noted the work of the Chinese between Port Moody and New Westminster: 'These human bees are always employed and you can see from your carriage windows innumerable little wooden railroads, which they have constructed, on which they take the cordwood by horse power out of the forest to the track where it is piled ... they are paid a dollar a cord by the Railway Company, and ... make two dollars a day to the hand. It is evident they intend to be permanently employed, for they are making substantial little houses along the track.'[19] However, by the turn of the century Japanese workers were increasingly active in taking out cordwood as well.[20]

Like all wage workers, loggers faced the problem of getting paid when companies disbanded, disappeared, or went broke. However, the problem was so great in the woods that in the 1880s the provincial government brought in legislation to deal specifically with loggers. Logging operators contracted with sawmills to deliver logs, but too often after the mills paid the logging-camp operators, wages for the workers were not forthcoming. In April 1888 dissatisfied loggers seized a boom of logs on the North Arm of Burrard Inlet, refusing to release it until they had received their wages.[21]

The provincial government moved to address the grievance of loggers, who, after spending four or five months in the woods, wanted their wages on demand. An 1888 piece of legislation stipulated that the sawmill which purchased logs had to ensure that the loggers working for a contractor had been paid before the mill paid the contractor and took possession of the logs. The logging contractor was required to provide payroll sheets as evidence. The sheets showed the names of the workers, the number of days employed, the rate of pay, and the total amount due, as well as the signature of each man employed to prove that he had been paid in full. If this procedure was not followed, then the purchaser of the logs, the sawmill company, would be held liable for the unpaid wages of the workmen. The act was weak. In the fall of 1888, Norman Stewart, a logging contractor, engaged with Royal City Planing Mills to sell them a fixed number of shingle bolts. He hired a number of men, but when he was paid by the mill company, he did not pay his workers. One of the workers, Daniel White, took Royal City Planing Mills to court, but the judge ruled that the legislation applied only to logs and that shingle bolts were not covered. In August 1890, C.W. Murray closed his logging camp on Jervis Inlet. The camp had been producing logs for Hastings Saw Mill, owned by British Columbia Mill Trading & Timber Company. The camp workers were not paid and brought a suit against the sawmill company. The judge dismissed the case, 'giving it as his opinion that the action would only be in a case where the mill company owned the limit and hired the men to get out the logs.' The Woodman's Lien for Wages Act in 1895 attempted to enhance the claims of workers, but getting paid continued to be a problem.[22]

Injury and death were ever-present features of labour in the woods. There are no convenient statistics on injuries and deaths in the nineteenth century to enable these gruesome realities to be quantified neatly and cleanly. However, Vancouver and New Westminster newspapers saw these injuries as news until the 1890s, and though the record is not complete and incapable of showing statistical trends, we can make some comments on the slaughter in the woods. But workers in the woods would likely have found little solace in statistical data because all workers faced death daily. The threat of an industrial accident was part of every working day, and this is the most striking aspect of the historical record.

Logging and sawmilling were not noted as particularly dangerous in the nineteenth century; almost all male working-class jobs entailed risks. Agriculture, building railways, running railways, mining, factory production, and fishing all put workers in dangerous situations. Interacting with moving parts in machines, and manipulating large, heavy objects, were at the core of the work world. Coping with injury and pain were related to maleness as well as industrial work. Recent scholarly studies of the historical construction of masculinity note the importance of physical strength and the ability to cope with dangerous situations as hallmarks of manliness.[23]

Logging accidents were reported as early as the 1860s. George Hazelwood was working in the Moodyville logging camp on Burrard Inlet on 17 June 1867 when a large log rolled over him. He was taken to the Royal Columbian hospital in New Westminster where one finger and a leg, with a badly crushed foot and fractures in two places, were amputated. Hazelwood fell into a coma on 10 July and died ten days later. He was thirty-eight years old, a native of Yorkshire, England.[24] The increase in the number of loggers in the camps after 1885 meant that there were more accidents. In 1889 and 1890 the newspapers reported six logging deaths and five serious injuries. Death and injury in the woods continued in the twentieth century. In the decade after 1909, 123 loggers died in industrial accidents.[25]

Young and old had to be on the alert. William Hamilton, a twenty-three-year-old from New Brunswick, was killed by a falling tree at Charleston's logging camp, near the McInnes farm on the Fraser River, on 24 August 1888. In August 1890, James Marks, aged seventy, was injured at Taylor's camp near Hastings: a commentator noted that 'he was especially vigorous for a man of his age.'[26] Nor was any ethnic or racial group immune. In August 1865 a nameless Chinese worker cutting timber near New Westminster was killed instantly when a tree fell on him.[27] In July 1882, Augustus Anderson, a Swede who was about twenty-four years old, was crushed to death between two great logs while yarding at William McGee's camp at Mud Bay.[28] An unnamed Native logger had three toes chopped off at a Howe Sound camp on 9 May 1889: 'he did not think it necessary to come to the city, but attended to his injured members after the aboriginal habit.'[29]

Nor did experience shield workers from death in the woods. In May 1890, well-known logger Charles L. Daggett was killed. The fifty-three-year-old single man was from Maine, but had been logging in the area for twenty-five years. In 1890 he was the camp foreman at Hayden's camp on Gambier Island: 'in a moment of thoughtlessness, he stepped in between two huge logs six feet in diameter. The log he was hauling rolled against the outside one, and immediately he was caught about the thighs and fearfully crushed. He called for assistance, but being in the habit of shouting to the oxen no one took any notice of it. He was 15 minutes in this position and only lived two hours after he was rescued.'[30] Jim McKim was a popular, experienced logger from Nova Scotia known as the 'logger king.' He was forty years old, had worked in coastal logging for nine years, and on 29 September 1894 operated a camp on Howe Sound. McKim was cutting some tobacco, standing alongside the chute down which logs were being run, when a log jumped from the chute, striking the man and causing instant death. The inquest concluded that McKim, despite his experience, had been careless:

> The chute in use was one having cross skids. There was considerable dry timber cut. This as a rule would not run at all well on a dry day. Saturday was wet, and advantage was taken of the fact to run these dry logs. No dirt was put on the chute to make them run slow as was done when green logs were being run and they made the average trip of the chute in 30 seconds. As the chute was half a mile long they must have travelled at a very rapid pace. Other logs had jumped the chute, and Mr. McKim was taking a risk in running them as he did.

McKim's death was the third coastal logging fatality in two weeks.[31]

Most logging injuries and deaths occurred when large timbers rolled onto workers, either when being felled or yarded. Animal and steam power both held their perils. In October 1883, Henry Chantrall was logging with a team of oxen at Mud Bay, when the team ran away: 'the chain hook caught Mr. Chantrall by the foot, and he was dragged a considerable distance, and had his leg broken above the ankle.'[32] A. Morrison was gored by an ox at a northern coastal camp in June 1889, 'his abdomen being ripped up by the angry animal.'[33] A government official noted a logging accident involving a steam-donkey engineer in 1907. He was operating without a certificate and worked at the Gulf Lumber Company's camp on Jervis Inlet: 'The throttle valve on the logging engine was said to be leaking, and, in order to haul off some line, the engineer stood up on the side of the bull-wheel. The engine made one strike and jammed his foot between the wheel and frame, and in his excitement he reached for the throttle to close it, pulled it too far and opened up the steam port, starting the engine up and taking his leg off above the knee before assistance arrived.'[34] The pace of work in a steam camp, however, likely increased the rate of accidents.

As the logging camps stretched up the coast after 1885, work sites were farther and farther from medical facilities. An industrial accident in May 1892 graphically illustrates the predicament of injured workers who were in camps far from medical treatment:

> Jack Sinclair, foreman in Don Mathison's camp on Hardwicke Island, was sniping a log on Thursday morning when another log rolled up against it, wedging in his leg so tightly that it was found impossible to move the log. His fellow workmen were forced to chop away part of the log in order to release the imprisoned limb. Sinclair's agony while this work was in progress must have been terrible, as every stroke of the axes upon the log would send a shock through the mangled leg. When it was got out the calf was all crushed and two deep wounds, eight to ten inches long, had been inflicted on the upper part of the leg. The *Comox* had left the evening before, and as there was no prospect of another steamer for several days it was decided to bring the injured man down in a row boat. Capt. Myers, Dick Flowers, Ben. Clegg, and Alex Macdonald accordingly started with him as soon as arrangements for his comfort could be made. They left at 9:30 on Thursday morning and arrived here [Vancouver] at 9:30 this morning [Saturday] showing that they must have rowed hard and steadily.[35]

In the spring of 1893, after two serious logging accidents, a Vancouver newspaper noted that logging had become as dangerous as railroading and that steps should be taken: 'The distance that these poor men have to come for medical treatment adds to their suffering. The way in which one of the arms was tied up necessitated a great deal more pain in having it undone and dressed than would have been caused if it been properly attended to, and if while in the city in the winter the loggers will form a class, *The World* knows of a surgeon who will willingly give them a few addresses on first aid to the wounded.'[36] In 1895 two fatalities at one logging camp in one month prompted the coroner to suggest that there was gross carelessness on the part of someone. He added that, because of the great number of logging fatalities, a royal commission should be held into the system of logging on the coast.[37]

The investigation of the logging system never took place and it was a decade before medical facilities were established in the northern coastal region. In the meantime loggers participated in a program that at least provided insurance for medical treatment. It began at St Mary's hospital, which opened in New Westminster in 1887. The Catholic sisters travelled to the northern camps by steamer, collecting money from the loggers. In 1892 the fee was $10 per year. In return for their contribution, loggers were guaranteed free medical treatment at St Mary's. The hospital was referred to as the 'haven for injured lumbermen,' and in the early 1890s most injured loggers from up the coast travelled beyond Vancouver to be treated in New Westminster.[38]

There was a more substantial response to the injuries in the woods in the next decade. In June 1904, Reverend John Antle of the Columbia Coat Mission made the journey from Vancouver to Alert Bay, calling at settlements and logging camps along the way. He found some 3,000 men along the coast, chiefly at logging camps, 'cut off from the comforts of civilization, destitute of medical and surgical aid, and almost entirely lacking the influence of religion.' He resolved to equip a small steamer as a hospital and dispensary to travel the coast, offering medical services as well as books, magazines, and religious services. The *Columbia* was launched in April 1905. British Columbia Mills Timber & Trading, after meeting Antle, built a hospital at Rock Bay, which opened in July 1905. The hospital was furnished by the Victorian Order of Nurses and operated by the Columbia Coast Mission. Two more hospitals were built up the coast, one on Texada Island and the other at Alert Bay. In 1909 the Columbia Coast Mission, which had the official support of the B.C. Loggers Association, treated 1,837 cases, 549 of these on board the *Columbia,* including 240 operations, 1,475 surgical dressings, and 1,058 prescriptions. In 1911 the Mission operated a second vessel, *Columbia II,* and operated the three hospitals. Loggers contributed to the maintenance of the services: the Hastings Saw Mill Company deducted one dollar per month from workers' pay cheques as a medical fee.[39]

A particular lifestyle came to be associated with loggers around the turn of the century. During layoffs and shutdowns due to rain or snow, the loggers congregated in the city for relaxation and recuperation. Loggers congregated in their haunts in Vancouver, engaging in boisterous activities, often under the critical eye of more 'respectable' citizens.[40] Already in the nineteenth century alcohol consumption was a noted aspect of logger culture. In 1884 an exasperated foreman at a Dominion Sawmill Co. logging camp on Campbell Creek complained to management that when the company called a man to New Westminster on company business, the man often went on a drinking spree: 'The timber is not so good as I thought[.] we leave about half in the woods[.] there is only 5 men here and when any of them goes to town how can you expect that we can do anything when you keep them in town for over a week[.] if I find out they are on a drunk and got settled satisfactory at the office I will send them back[.] so if you have to keep a man waiting for any length of time you better send another in his place[.]'[41] A Vancouver newspaperman saw the loggers more positively in October 1892: 'Lumbermen are crowding into the city and as usual are making things lively. No class of men spend money more freely than the loggers, and none get more fun out of it.'[42] Alcohol was present in the camps. Boss logger Ward G. DeBeck, who was logging for Haslam's Nanaimo Sawmill at Discovery Pass on the east coast of Vancouver Island in 1889, wanted the government to employ a police boat to cruise the coast and crack down on the sloops trading in alcohol: 'The very worst sort of fire-water is supplied to Indians and to the logging camps, and the trade is constantly increasing.'[43]

In the Fraser Valley, loggers were more closely integrated into the rural communities. Some farmers, such as the Belmont orchardist A. Matheson, worked in logging camps near their home. Others, such as the Port Haney farmer George Blackstock, the Langley farmer Francis Joseph Livingstone, and the Ladner's Landing farmer John Jackson Burr, both farmed and operated logging camps. The logging and farming communities in the Fraser Valley were close. The Loggers Ball, put on by the men employed at the Royal City Planing Mills camp in 1891, was held at the Surrey Town Hall Centre and attracted nearly 130 couples. Dancing began at 9:00 p.m., supper was served at midnight, and then dancing continued until 5:00 a.m. Church services, too, were held in Fraser Valley camps, led by a minister coming out from New Westminster.[44]

After 1900, employers and social reformers became increasingly concerned with the lifestyle of loggers in Vancouver. Loggers were characterized as careless, gullible, and shiftless, making no provision for their futures. When in Vancouver they frequented unhealthy establishments, were prey to confidence men, socialized with prostitutes, and squandered money in saloons. This had implications for the stability of the labour force and the efficient operation of the production system:

> The better the wages the shorter time they will work, because it takes them a shorter time to lay by enough money to have a time in the city. There must be a reason for this, and no doubt the reason is mainly that although they work hard while they are working, their reward is great, and as they lead a monotonous hum-drum life in the woods, the temptation to make up for innumerable hardships when in the tall timbers is too irresistible to withstand, when they draw down as they do large sums of money, sometimes running into hundreds of dollars.[45]

In 1908, W.I. Paterson, a logging operator, appealed to the churches to take on a greater role 'in making the Lumberjack of British Columbia a better citizen,' arguing that 'the dissatisfaction of working men was caused by drink,' and declared that this was because nobody cared for them in the camps and towns.[46] By 1910 the Canadian Camp Brotherhood, funded by businesses and churches, was active in coastal camps, and at their headquarters in Vancouver on Powell Street, in the heart of the area where many loggers resided when in town, the association engaged in intellectual, physical, and moral work to improve the lives of loggers. The Vancouver Sailors' and Loggers' Society, also called the Strathcona Institute, provided a home for sailors and loggers who were out of work. In 1911, the institute raised money for the construction of a new, larger building with 150 rooms, as well as a hospital emergency ward, restaurant, lecture hall, and recreation, writing, and smoking rooms. Religious services and help in finding work were also provided.[47]

Alongside this negative image of the logger, however, was a much more complimentary perspective. Lumber trade journals, which emerged specifically to deal with British Columbia around the turn of the century, began to construct a written history of the coastal lumber industry and articulate a vision of loggers and logging. These journals, full of advertisements for equipment, machinery, and supplies, were geared towards owners and managers, men working in offices. The vision of the loggers portrayed in the journals reinforced the masculinity of the office workers, some of whom had previously worked in the woods, and at least gave them the sense of participating in an industry where manly virtues prevailed. Loggers were not effete slaves to their desks and paper shuffling. Loggers were hardworking; curiously, according to R.H. Alexander, it was often the best and hardest workers who engaged in the most raucous drinking sprees.[48] Loggers were physically strong, worked in the great outdoors physically manipulating giant logs, ate heartily due to their extreme physical exertion, and were imbued with the special 'character which the forest gives to men.'[49] City slickers had little conception of the difficulties and joys of work in the woods:

> Let me tell you, my friends in the lawn tennis suits an' the cute little red ribbon lids,
> To the man in the woods in his snowpacks or boots you are nothin' but sissies an' kids.[50]

The trade journals also celebrated the particular jargon of loggers, the unique and colourful language, understood only by workers and those familiar with the industry, used to define things and processes in the woods, reinforcing the notion of loggers as a breed apart. Logging was not a mere a job: 'There is not a logger that will not say that life in the woods is the hardest of all callings and most fraught with dangers to life and limb, and still where is there a logger, as long as his youth and strength lasts, who will not, after a short season in town, hanker for the woods again?'[51] The management trade journals omitted the fact that logging was an industrial job, ignoring wage cuts, monotony, unemployment, cruel bosses, and the pain of death and injury, but did reflect that the transient life which took loggers back and forth between camps and the city had its appeal.

Despite these split images, there was an increasing focus on changing loggers' lives in the first decade of the century. A controversial organization spearheaded a drive to improve camp conditions in coastal British Columbia in 1910 and 1911. The United Brotherhood of America, known informally as the Loggers' Union, was dedicated to combining the good points of trade unions, medical associations, and fraternal societies. They planned to operate an employment bureau, and provide free medical and legal services. The goal was harmony between employers and employees, and apparently they had the support of some employers. The Vancouver Trades and Labour Council was suspicious of the brotherhood, because it invaded union jurisdiction claimed by another union and because of its commitment not to interfere with hours of work and wages. It was even alleged that the brotherhood was a money-making scheme for the promoters.[52] The extent of the organization is unclear, but it had enough clout that its president and secretary were received by Premier McBride in 1911.

The organization focused on three issues, issues that would also be prominent in the rise of union activism among loggers after the First World War. They wanted to be paid in cash when convenient, and, if this was not possible, they wanted at least enough cash to pay for their transportation to the nearest town. They also wanted a full medical chest in each camp, as well as a boat to quickly transport injured workers to a hospital or doctor. The final issue, and the one which was pushed most vociferously, was for improved living conditions in the camps. The list of demands included: '(1) bedding owned by the company be made subject to sanitary inspection, dealt out to each man on joining the camp and charged to him, with rebatement of cost of same on his departure; (2) bath houses and wash houses to be at least 50 feet from bunk house, and to be of modern construction ...; (3) modern and sanitary bunks with springs and mattresses, and comfortable, well ventilated quarters for the men ...' The union also wanted the government to enforce the Health Act provisions that applied to work camps.[53]

This focus on domestic considerations, rather than wages or hours of work, is noteworthy. There was no appeal to any good old days, suggesting that there was no deterioration of living standards in the camps, that they always had been poor or basic. Likely organizers chose this issue for tactical reasons. Poor camp conditions were the norm in the coastal region and the issue reflected a much broader aspect of the Christian-based reform movement in Canada. Camp life, and bachelor life in particular, were portrayed as debilitating and evil. As reformer J.S. Woodsworth wrote in 1909: 'It is always a disaster for men to congregate together, whether for a longer or shorter period without the blessed influences of a home in which there are women and children. This is true of thousands who spend their years in the lumber-camps, and in the mines of the far North and West.'[54] In the 1940s a sympathetic historian of workers in the North U.S. logging industry commented on pre–First World War loggers, who were migratory, without wives and families, and living in filthy camps, and who participated for brief periods in an unhealthy urban social life dominated by alcohol consumption and prostitution: 'The leading of unnatural lives in such abnormal surroundings, with no way of avoiding them, was bound to have definite effects on the workers.'[55]

That this view comes to dominate in the years after 1900 is clear, but the turn to domesticity was forged by organizers, activists, and broader social ideas about men and society. The greater number of men working in coastal camps after 1900 and their more noticeable presence in disreputable areas of Vancouver raised concerns. Whether coastal loggers at the turn of the century believed that their lives were unnatural and their surroundings abnormal, or whether indeed they were, is questionable. Society's perceptions of loggers and loggers' perceptions of themselves had to be worked out in a changing context.

Debates about living conditions and unionization were intertwined with notions of masculinity. Ideals of what constituted a logger were contradictory. Were loggers special men who endured conditions that were unacceptable in other walks of life or did they deserve basic domestic amenities enjoyed by most others in society? Did real men embrace adverse conditions without whining, independently dealing with employers and moving to a different camp if conditions did not meet their standards, or did real men act collectively with their fellow workers in a protest against unacceptable situations? Union activity in the camps prior to 1913 was fleeting, but after 1917 a more effective loggers' union was organized. Loggers and activists were able to develop a persuasive discourse that linked economic exploitation, a vision of masculinity based in collective action, and a sense of human dignity reflected in healthy, sanitary camps.[56]

There is scattered evidence of organized job action by loggers prior to 1913. In 1900 a few loggers went on strike at W.L. Tait's camp on the North Arm of the Fraser River, to protest the hiring of Chinese workers.[57] The next year the Victoria Lumber & Manufacturing Company began running their own logging camps at Chemainus instead of relying on contractors. As part of the transition they introduced a new wage scale, which included a pay cut for 35 of the 110 men. The disaffected men struck on 16 May, but the company hired new workers, and the strikers were forced to find work elsewhere. The job action coincided with the company's first hiring of Japanese workers in the woods, which may have been part of the dispute as well.[58] In 1905 there was another strike in a Victoria Lumber & Manufacturing camp. The men were protesting a reduction in work time from ten to nine hours, which lessened their pay.[59]

Part of the reason for the weak organization by loggers was the coordinated response of employers. Logging companies attempted to rationalize wage schedules and control the fluctuations in the labour supply. By January 1904 the British Columbia Loggers' Association published a wage scale for logging-camp labour in an attempt to set industry-wide rates and discourage companies from driving up wages by competing for labour. In the spring of 1906 the association established its own employment bureau to recruit labour for the woods. The traditional tours of hotels, bars, and private employment agencies were not providing the necessary workers.

In 1907, due to a labour shortage, the association advertised across Canada for loggers. They recruited 150 workers from Winnipeg and some from Quebec. The tight labour market in the Far West meant that the companies, to their chagrin, had to hire many inexperienced loggers at premium wages: 'They are now, as a class, overpaid for their labor. They have the advantage of being enormously in demand, and as a result there are numerous green hands among them, who get the same wages as the skilled ones.'[60] A downturn in markets imposed a rough discipline on the loggers; over the course of 1907, log prices dropped from $12 per thousand board feet to $8, wages were cut, and in December 1908 some 500 unemployed men marched on city hall in Vancouver, asking council for employment.[61]

Coastal loggers in 1913 remained hostage to fluctuating markets and the controls of employers. They had no union organization and were still unable to mobilize strike action to further their goals. The workforce was largely white, but the job hierarchy and the isolation of camps dispersed throughout the region made organization difficult. By the first decade of the twentieth century, too, competing discourses that addressed the connection between the job culture and notions of masculinity divided workers and undermined collective activity.

9

Millworkers

By the turn of the twentieth century, the defining features of the mill workforce in coastal British Columbia were race and skill, and the two were intertwined. White workers held the best jobs and made the most money, while Asians worked at the more disagreeable jobs, had little opportunity for advancement, and earned less in wages. While the racial division became pronounced after 1885, the job hierarchy based on skill and wage was present since the 1860s.[1]

In the sawmills, there were a number of skilled jobs that were well paid. The sawyer had a particularly important job. A round log that tapered from one end to the other entered the mill, and the sawyer had to decide how to make the initial saw cuts to maximize the amount of square, dimension lumber that could be produced from each log. These were very important decisions, as logs cost money and a mill's profits depended on getting the best value in square lumber from each log. The saw filer, who maintained and repaired the saw blades, was another well-paid, skilled worker. The smashing of a circular saw blade due to poor repair shut down production in a mill, at great cost to the employer. Tallymen kept track of the logs and lumber on the mill site, and were connected to the shipping department. They were literate and capable with figures; they did not engage in manual work or the operation of machines, and the job was often a starting point in an occupational ladder that led to a management position. Benjamin Springer began as a tallyman at the Moodyville sawmill, became the bookkeeper, and in 1882 was promoted to the position of manager of the whole sawmill operation.[2]

A second tier of millworkers, including edgermen and trimmermen, operated saws that further refined the dimensions of a piece of lumber.

The sawmill workforce also included workers that were not directly involved with logs and lumber: blacksmiths, machinists, carpenters, and millwrights did construction and repairs, and steam engineers kept up the power in steam engines. Finally, there was a host of unskilled workers in the mills, physically turning or canting logs on carriages to position them for runs through the saw, moving and piling lumber in the mill and mill yard, and transferring lumber into ships from the wharves of the export mills. The mill workforce, then, was built around a core of skilled, well-paid workers, with a supporting staff of replaceable workers known as ordinary labourers. The mill wage scale in 1894 at Royal City Planing Mills in New Westminster shows the job hierarchy based on skill. The sawyer was paid $75 per month; the resawyer $55; the setter, dogger, first fireman, and shingle sawyer $45; the marker, slab sawyer, truckman, second fireman, and slab carrier between $36 and $40; and yard hands $34.[3]

Data from the 1881 and 1891 census data for the New Westminster Census District reveal much about the mill workforce.[4] Strikingly, over the course of the decade, Chinese workers assumed a significant number of jobs in the mills. In 1881, of 122 millworkers identified, 31.1 per cent (38) were born in the British Isles. A further 7.3 per cent (9) were born in Ontario, 6.6 per cent (8) in the United States, 5.7 per cent (7) in British Columbia, 4.9 per cent (6) in Quebec and Nova Scotia each, and 2.5 per cent (3) in New Brunswick. These British, Canadian, and U.S. workers constituted 63 per cent of the millworkers. A further 18.1 per cent was made up of workers from the Sandwich Islands, Spain, Russia, and Chile; and a smattering of workers born in Finland, Norway, Sweden, France, and the West Indies rounded out the identifiable labour force. In 1891, of the 590 sawmill workers enumerated in the census data, fully 208, or 35.3 per cent, were Chinese. The median age of the Chinese labourers was thirty-three, and 20.2 per cent (42) were married. At the same time, the non-Chinese workforce was becoming more British and Canadian. Of the non-Chinese workers, 31.7 per cent (121) were born in the British Isles, 21.2 per cent (81) in Ontario, 7.6 per cent (29) in Quebec, and 22.5 per cent (86) in other parts of Canada. These men constituted 83 per cent of the non-Chinese sawmill workforce. A further 6.8 per cent (26) were born in the United States, and 1.5 per cent (6) in Newfoundland.

Age and marital-status information regarding the non-Chinese workers is also revealing. There was a notable difference between skilled and unskilled workers. Of the 122 millworkers identified in 1881, 15 were skilled workers, such as saw filers, machinists, blacksmiths, millwrights, sawyers, tallymen, engineers, and watchmen. Their median age was forty-two, making them older as a group than unskilled workers, who had a median age of thirty-two. Some 20 per cent of the skilled and 14.8 per cent of the unskilled were married. In 1891, looking at the non-Chinese workers again, there were 252 unskilled workers and 130 skilled workers. The median age of the skilled was thirty-one, as opposed to twenty-six among the unskilled, and the skilled were more likely to be married (34.6 per cent, as compared with 16.3 per cent of the unskilled). Skilled workers were older and more likely to be married, suggesting that they were more settled and stayed at their jobs longer. The skilled workers were also more concentrated ethnically: in 1881 all the skilled workers, with the exception of one German, were born in the British Isles or Canada. Skill, age, ethnicity, and marital rates were intertwined. A core of skilled workers of similar descent ran the mills, supported by an ethnically diverse, younger workforce that was increasingly made up of Chinese workers as the 1880s progressed. While the census offers insights into the makeup of the sawmill workforce, revealing general trends, it is hardly a perfect source. The range of races and ethnic groups represented in the export mills on Burrard Inlet was extensive.[5]

Transiency was a notable feature of the mill labour force, in part reflecting the fact that the unskilled were overwhelmingly single and able to easily move to other places and jobs, which at least provided a change of environment. In 1874 Lieutenant-Colonel Houghton established a militia unit in New Westminster, which had a more diverse, larger, and more stable population than Burrard Inlet, where it was impossible to set up a unit: 'the migratory character of the population rendered it unable to attempt creation of a corps there.'[6] The availability of jobs fluctuated with the lumber economy. In good times, such as 1872 and 1882, there were plenty of jobs and premium wages, while in economic downturns there were layoffs and wage reductions. Likely, too, transiency reflected the fact that it was difficult for young men to get skilled positions. The people who had the jobs stayed, and when new positions opened up they were filled by men who had acquired the necessary training elsewhere, and apprenticeships went to sons of skilled workers.

For the unskilled, as well as lack of opportunity, the tediousness of the work also contributed to job turnover. The nature of mill work was expressed by a novelist who experienced the 11.5-hour workday in a New Westminster sawmill in the 1880s:

> In ... the half-open mill, one storey up in the air, I passed the days, with the whirr of belts above and below, the scream of the circular saws as they bit the advancing log of pine or spruce or Douglas fir, with the strips of bitten-out wood thrown out in a stream, and clouds of smaller sawdust, with the smiting of mallets on wedges in the cut, and the heavy fall on the greasy skids of the divided tree. And then, in the pool below, stood a long figure with a pole balancing on a round log, pushing it into place, then the hammer driving in iron clamps or dogs, and the chain, revolving on the drum, dragging the ponderous tree to the saw, and then its rolling over and over on to the carriage, and afterwards more saw-screaming and sawdust and wedge-driving. So hour after hour, till the trees, rude and huge, fall into planks and boards and squared timbers – large for bridges or small for posts or pickets, and the waste cut into laths, and the sawdust burning in the gaping furnaces to drive the saw again.[7]

The presence of First Nations people in the mill workforce deserves further comment. Native peoples did participate in mill work, but their labour was not crucial to the lumbering economy. By 1864 the milling and logging operation at the Alberni depended on the labour of about 270 white workers, three-quarters of whom were Canadians and Americans, and a few local First Nations people in ancillary, unskilled positions.[8] Gilbert Malcolm Sproat, manager of the operation, was not reliant on Native labour: 'Some of the Indians became fair workmen, and their labour was worth half-a-dollar a day and rations, or about one-third the value of an ordinary white labourer's work; but, on the whole, I found that Indians were unprofitable workmen. They make better sailors than labourers ...'[9] At Burrard Inlet in the 1870s First Nations people worked inside the sawmill operation at Moodyville, which was adjacent to a Native community, and others worked as stevedores in the 1870s.[10] In 1884 the workforce at the Moodyville sawmill was described as being made up of seventy-five white men, besides a force of 'Indians and Chinamen,' and likewise the Hastings mill employed sixty white workers plus a number of Native and Chinese workers.[11]

The 1891 census clearly reveals the presence of First Nations workers in the Moodyville sawmill. Fifteen Native workers toiled inside the mill, but only one of these was married. Married Native people were more likely to work as stevedores, where family crews worked together as a unit. Of the thirty-three dock workers at Moodyville, twenty-six were Native people. Of these, twenty-three were married. Further,

Table 9.1 Asian and White Workers in Selected Mills, 1901

	Whites	Chinese	Japanese
Victoria Lumber & Mfg	58	56	56
Hastings Saw Mill	164	0	93
Royal City Planing Mills (New Westminster mill)	180	57	29
Moodyville Sawmills	60	10	40
Sayward Mills (Victoria)	30–40	60–70	0
Haslam Mill (Nanaimo)	39	13	9
North Pacific Lumber Co.	45	0	46
Robertson & Hackett	80	0	20
Royal City Planing Mills (Vancouver mill)	90	11	60
Brunette Saw Mills	168	10	78
Pacific Coast Co. (nine shingle mills)	210	105	?
Spicer & Co. (shingle mill)	20	30	5
McNair (shingle mill)	159	27	42
Heaps & Co. (shingle mill)	56	21	27

Source: Canada, 'Report of the Commissioners Appointed to Inquire into the Subject of Chinese and Japanese Immigration into the Province of British Columbia,' *Sessional Papers, 1902*, no. 36, vol. 13, Sessional Paper 54 (Ottawa: King's Printer, 1902), 127, 360.

seven were women, and the ages of the workers ranged from fourteen to eighty-five. The seven non-Native longshoremen identified at Moodyville in the census were also likely connected to the Native community: at least two were 'Kanakas,' and one English-born worker was married to a Native woman.[12] The longshoring crews worked under a contract system, and a number of mills employed stevedore companies, which hired labour of various ethnic backgrounds, to manage the loading of ships.

For First Nations people and the white population of the province, the increasing number of Asian workers in the mills after 1885 was alarming. In the late 1880s and through the 1890s, Japanese migrants, as well as Chinese, found employment in coastal mills. In 1901 a federal royal commission investigated Chinese and Japanese immigration in British Columbia, and the mill workforce was a major focus of the study. Information from a number of mills gave a sense of the racial hiring policies (see table 9.1).

The commissioners concluded that over half the workers in the shingle industry and a significant proportion in the sawmills were Asian. After

1906, migrants from India arrived in British Columbia, and they, too, were integrated into the mill labour force. A survey of the major sawmills in Vancouver in early 1908 located 2,443 workers, comprising 1,067 whites, 802 Japanese, 399 Chinese, and 175 East Indians.[13] However, in the minds of many British Columbians, Chinese, Japanese, and Sikhs had taken over in the milling industry, allowing no opportunities for white or Native workers.[14]

Inside the mills, attitudes regarding race and skill were intertwined in allocating jobs and establishing pay scales. Employers were quite clear on the value of various groups in the production process: white workers and Native Indians were preferred, although Native workers were considered somewhat unreliable; Japanese and East Indian workers were capable and able to handle many jobs done by whites; and Chinese workers were the least efficient at most jobs. The best-paid jobs operating machines were the preserve of white labour, although some Japanese workers ran saws. Chinese workers were relegated to unskilled labouring positions. In a New Westminster mill in 1904, the head sawyer was paid $4.00–$5.00 per day, and the carriage riders, edgermen, and trimmers were paid $2.00–$2.75. All of these were white workers, as were the well-paid carpenters, blacksmiths, tallymen, and steam engineers. The unskilled labourers in the mill were paid along racial lines: white labour received from $1.75 to $2.00 per day, Japanese labour from $1.00 to $1.60, and Chinese workers $0.90 to $1.30. In the shingle mill the white sawyers received from $0.11 to $0.15 per thousand shingles, and the Chinese packers half the amount paid to sawyers.[15] As the decade progressed, Chinese workers increasingly worked as sawyers in the shingle mills, and Japanese workers operated saws in coastal sawmills.

For mill owners who competed with the producers on Puget Sound, the employment of low-wage Asians gave them an advantage. Restrictions on Chinese immigration into the United States since 1882 and low migration rates from Japan meant that the Puget Sound millworkers were overwhelmingly white. In 1901 only one large mill operation in Washington state employed Japanese workers. In waterborne export markets, where British Columbia faced U.S. competition, and in the U.S. shingle market where British Columbia producers were active, the Asian workers were seen by the coastal mill owners as necessary in maintaining their trade. And since the export mills and shingle mills also produced for the Canadian market, producers who focused on the domestic trade felt forced to hire low-wage Asians to sustain their businesses.

Employers also liked the reliability of Asian workers. Mill owners hired gangs of Asian workers through a contractor, and if a man was missing from work the contractor would ensure that another man would be on the job. The Thurston–Flavelle operation employed some 125 workers in 1912, including 80 Chinese, 4 or 5 Japanese, and 6 Sikhs. Wages for whites averaged $2.25 for the ten-hour day, $1.95 for Japanese and Sikh workers, and $1.85 for the Chinese. Jai Tai was the Chinese labour contractor for the sawmill and he also ran a boarding house and a store for the Chinese workers. The company made out a single monthly cheque to Jai Tai, and he was responsible for paying the Chinese sawmill workers. In the company's shingle mill, where all the sawyers and packers were Chinese, Soon Kee, of the Tai Sing Company of Vancouver, was in charge of supplying, managing, and paying workers. The Chinese sawmill workers and shingle millworkers were from different tongs, and a separate boarding house had to be constructed for the shingle crew. At this operation the Japanese and Sikh workers, who were known by numbers and not names, were given individual cheques.[16]

Although coastal employers increasingly felt that Chinese workers were most suitable for work as shingle packers and even sawyers, they almost all felt that white workers on the whole were more productive millworkers. Moreover, they generally supported restrictions on Asian immigration. As R.H. Alexander commented in 1901: 'I should far rather prefer to have white men as citizens. I would not like to see the Chinese and Japanese obtain the franchise. I would not like to see our country governed by them. I should not like to see any further immigration of them to enter into competition with white men. They would not adapt themselves to our political economy. There is no chance of their becoming citizens in the real sense of the term.'[17]

While white mill owners believed in restricted immigration and sincerely preferred to employ white labour, there was a rub. White labour would have to work for lower wages. At times mill owners replaced Asian workers in their operations with white workers. In 1906, the Victoria Lumber & Manufacturing Company began gradually replacing Chinese and Japanese workers with white workers; in 1909, Fraser Mills began bringing in French Canadian workers from Quebec to displace Asians; in 1912, Small & Bucklin announced that it would only hire whites; and, in 1914, the Canadian Pacific Lumber Company replaced its crew of Asian workers with white labour.[18] However, such changes in hiring policies were largely short-lived, and Asian workers remained prominent in the mills. In the clash between cultural imperatives and the dictates of the market, the mill operators acquiesced to the values of profit and loss.

White workers in the mills benefited from the employment of Asian workers. The lower wages paid to Asian labour enabled skilled white workers to retain their jobs in a competitive environment and sustained their wages at a higher level than would otherwise have been possible. During economic downturns Asian workers often faced wage cuts, while the white workers were unaffected. For example, on 1 November 1907, at a time when many mills were shutting down due to a weak market, a New Westminster mill reduced the wages of Chinese, Japanese, and East Indian workers from $1.75 to $1.40 per day in order to keep operating, and it was expected that other mills would also 'reduce the wages paid to this class of labour.'[19] Asians, then, endured the brunt of the slack times, while white workers retained their wage levels. According to a prominent mill manager, 'the cheap wages of the Oriental in many instances enables the employment of a proportion of white labor that would not be employed. They take the place of the style of labor like that in the manufactories in Europe that's done by women and children at low wages.'[20] Asian millworkers also were susceptible to seasonal fluctuations in other industries that employed Asians. In late September 1905, for example, the Lumber Branch of the British Columbia Lumber and Shingle Manufacturers' Association voted to reduce the wages of Asian millworkers because the end of the fishing season meant that many Asian workers were suddenly available.[21] Wages for white workers were unchanged.

White workers contributed to the process of racialization in coastal mills. As particular jobs were taken on by Asians, whites deemed these jobs as suitable only for Asians, refusing to perform them. From the perspective of whites, there developed a stigma associated with many jobs in the mills. R.H. Alexander remembered from the time when Asians were first employed in the mills: 'there's lots of jobs in the mill that isn't a question of wages. The [white] men won't work at it. He won't stick at it. I have had them working the pile driver. It was an unpleasant job ... We were paying these men $3.00 to $3.50 but they wouldn't do the work at that. They said, "Get the Jap to do that work." And they lay off whilst the Japs chopped ... so they could get at the head of the pile.'[22] Thomas Turnbull, a New Westminster labour leader, made the same point in 1909: 'Where the Oriental has control [of particular jobs] it has been the development of twenty-five or thirty years, with the result that men who would go in the business refuse to work alongside of Orientals. The White man who gets a job has to go and work with Orientals. White men object to that and it discourages them coming with the result they cannot get these men.'[23] A few years later, Turnbull noted that only a particular class of whites was willing to work with Asians in the mills, 'the kind of man that works for a few days and gets out as soon as he has a few dollars ... In the big mills you will see very little else but Orientals[,] just a few white men, foremen, and that sort of thing.'[24] Prejudice, as well as wage rates, discouraged whites from taking labouring jobs in the mills.

Asian workers took over certain unskilled jobs in the mills, and with time these jobs were labelled by white society as Asian jobs, allowing employers to maintain lower wages. Since Asians came to control the entry positions in the mill workforce, young white workers would not take the jobs for any length of time and would not learn the mill regimen in preparation for taking on more skilled jobs. This further entrenched Asians and afforded them the opportunity to take on more skilled jobs, such as operating saws. White workers were not getting the training. This was a common observation: 'Many [mill owners] stated that it was quite possible if there were no Orientals, steady white workers would be available, but they pointed out that in this case the scale of wages would be lower as in the States, where the ranks of skilled labor were continually being recruited from the ranks of unskilled labor, while in British Columbia, where Orientals were largely employed for unskilled labor there was no recruiting from the ranks of the unskilled ...'[25]

A Vancouver mill owner elaborated on the situation in 1910 in summing up his assessment of Asian labour in the mills. His comments illustrate the shifting line between what was considered white work and what was Asian work, and the ongoing historical process that reconfigured the labour process:

> the dearth of white labor is gradually throwing all classes of millwork into the hands of Chinese and Japs. The former are becoming excellent machine men – in fact I know a dozen or so as good as any white sawyers or planing machine men on the Coast. The Japs seem to be a bit afraid of the machines, and are employed mainly in the yards, but they are clever and adaptable and will no doubt become expert in time. In the old days we could hire a strong husky youth as fireman, and if he was any good it was not long before he qualified for the position of engineer. To-day we have to

> employ Chinese or Japs as firemen, with the result that white engineers are not being trained in the mills to replace those holding positions. Then again, we used to be able to hire boys as shingle packers; the smart ones quickly learned to handle the shingle machines, and some days the illness of the man at the edger saw gave the youth his chance for promotion. Another young fellow would watch the filer at work, practise a bit upon an old saw, and some fine morning a shingle saw 'trued' up by him would be found to cut well. Result, an assistant filer, who, in the course of six months or so, would be in charge of some filing room and in receipt of big pay. To-day very few new filers are being trained in British Columbia shingle mills because most of the packing is done by Orientals. Bad, isn't it? Most of the filers on the Coast are old hands, and we are badly in need of new blood from the East for this and other departments of our work.[26]

The process of racialization that developed after 1880 was complex, but still in 1913 prejudice from white employers and white workers ensured that Asian workers were not given the opportunity to take on the best-paid, most-skilled jobs in the mills. Racial divisions among Asians and between whites and Asians, as well as the transience of unskilled white labour, undermined the development of a cohesive mill community. British Columbia millworkers were a fragmented lot without overarching symbols of a job culture that gave definition and a sense of common purpose to their lives inside and outside the mills.

Although not as deadly as logging, labouring in the mills also held perils for workers. As well as falling prey to heavy logs and drowning in the boom pond, millworkers were endangered by moving machinery and sharp saw blades. A few examples drive home the ever-present dangers in the workplace.[27] On the morning of 24 September 1866 Isaac Steadman, an owner of the British Columbia Mill Company, was engaged as the sawyer: 'While in the act of using the gauging-stick one end of it came into contact with the saw while in motion, driving the other end of it through the groin, laying bare the femoral artery, and inflicting a most serious wound.'[28] Thomas Salmon was killed by an accident at the Hastings mill on 25 March 1880: he 'was working with the night crew, and while a log was being canted on the carriage it jumped over the blocks, and rolled down upon the unfortunate man, crushing his head so badly that he died from the effects, after lingering for about 48 hours.' On 17 June 1884 an unnamed First Nations worker at DeBeck's mill lost his arm; 'it came into collision with a revolving saw and disappeared like a flash.'[29]

At six o'clock on the evening of 22 November 1888, a Chinese worker at Royal City Planing Mills, New Westminster, lost four fingers while greasing the log carriage: 'The log was turned over and the unfortunate man happening to have one hand on the carriage at the time when the timber rolled over, the four fingers of his left hand were caught and ripped off. The sawyer who happened to see the accident fainted and fell, severely injuring his forehead.'[30] On 2 May 1892, Humphrey Cadwalleder was operating the resaw at Moodyville Sawmills: 'a large piece of timber was being canted onto the table preparatory to being resawn, when the chain carried it on the resaw, causing it to burst. One piece struck H. Cadwalleder on the back of the neck, inflicting a gash three inches long within an eighth of an inch of the jugular vein. He was also struck by another piece of the flying metal.' In early October 1892 Malcolm R. Wells was working at an emery wheel in a shingle mill on the North Arm of Burrard Inlet when the wheel burst: 'The flying fragments struck him in the face, breaking the bone of his nose, splintering the bone of his forehead and making a nasty gash.' Death occurred at Hastings Saw Mill on 17 September 1895. Robert McLeod was struck by a splinter thrown by the saw: 'The missile struck him in the groin, inflicting what was at once seen to be a very serious wound ... he died shortly before noon.' McLeod was the son-in-law of the Hastings Saw Mill clerk, and had recently moved to Vancouver in hopes that the coastal climate would benefit his wife's health. He was thirty years old and left two small children.[31]

Coping with pain and injury also allowed for the expression of courage, black humour, and masculine values. In May 1885, Milbert Matterson was injured at the Pioneer sawmill in Port Moody, 'having one finger sawed so that it hung by a thread, and three others partially off ... He had plenty of "sand" for when he saw the bloody hanging digit, he said to the assembled operatives, "Well, that's a fine-looking thing. Cut it off, boys."'[32]

Although the evidence is sketchy it appears that young, inexperienced men were particularly vulnerable to injury in the mills. On the morning of 31 May 1882, a young man named Glenn or Glen from Nova Scotia or Ontario started a new job at the DeBeck sawmill, working as an edgerman. Less than twenty minutes after starting work, his left hand 'was cut clean off about half way between the wrist and elbow!' At about 11:00 a.m. on the morning of 4 July 1882, the governor of the engine at the Royal City Planing Mill 'got out of order, by some means, throwing the entire machinery into confusion and causing a general smash up. A young man named William Doyle, who had been working on the boom, entered the mill, and, seeing that something was wrong, attempted to escape, but a fragment from a splintered pulley struck him with great force over the heart causing instant death.' Doyle, from Restigouche, New Brunswick, was about twenty years of age and had arrived on the coast about two weeks prior to his death. An elder brother worked at Angus Fraser's logging camp on the North Arm of the Fraser. On 14 September 1888 a boy named Wise, whose job it was to feed a bench saw at the Royal City Planing Mill manufactory, lost a finger.[33] On 18 July 1892, Hugh Reynolds, a twenty-year-old native of Ireland, drowned while loading shingle bolts from the water onto a car that ran them to the mill. It was his first day on the job. His brother was in charge of a ranch at Pitt Meadows.[34]

In some instances injured workers successfully sued employers for damages. John A. Foley, who was injured at Vancouver Sawmills in 1890, sued the owners of the mill, J.A. Webster and H.V. Edmonds, for $5,000. The jury found the owners negligent in failing to maintain the rollway and blocks in safe condition, awarding Foley full damages. In March 1893 a jury awarded Charles R. Scott $2,500, the equivalent of three years' wages, for the loss of his leg in an industrial accident in a suit against British Columbia Mills Timber & Trading Company under the Employers' Liability Act.[35] In most cases, however, employee negligence was deemed the cause of the accident and workers were left to fend for themselves.

Unlike loggers, millworkers engaged in job action from an early date. Strikes by sawmill workers revolved around wages, hours of work, and the increasing presence of Chinese workers in the mills. The first known strike in coastal sawmills took place in April 1886. Navvies, returning from inland railway construction camps, took jobs at Hastings Saw Mill, and with the support of the local Knights of Labor, a continent-wide workers' association, went on strike for shorter working hours and to ensure that no Chinese workers would be employed in the facility. This brief strike was successful in the short run. In October 1886, seven workers at the Leamy & Kyle sawmill on False Creek were discharged for refusing to work on Sundays. In the spring of 1887, pressure from the white community forced Royal City Planing Mills to discharge their Chinese workers.[36] A more substantial strike took place in June 1887. The employees at the Hastings Saw Mill, Moodyville Sawmills, Commercial Mills (Leamy & Kyle), and Royal City Planing Mills (False Creek) went on strike on 8 June 1887. They worked 11.5 hours per day and struck for the 10-hour workday. At the time of the strike, there were six vessels in the harbours loading lumber for foreign ports, and contracts stipulated that if the ships were not loaded in the agreed-upon time, then the mills would be forced to pay demurrage. Also, there were reports that a shipload of Chinese workers was due to arrive soon to engage in mill work. The Knights of Labor represented the strikers, and the mill owners met with a deputation of Knights. On 11 June 1887, the strike was settled, and the workers established the 10-hour workday in the sawmills.[37]

Chinese workers, though, remained in the mills. In February 1888, Royal City Planing Mills in New Westminster allowed the Chinese workers two days off to celebrate Chinese New Year. Pressure was applied at the political level to curb the Chinese presence. In 1890 the provincial legislature debated a bill to create British Columbia Mills Timber & Trading Company. Many in British Columbia wanted a clause inserted in the bill that would prevent Chinese workers from being employed in the company's plants. Henry Croft, MLA and lumberman, spoke out against including any anti-Chinese measures, arguing that the company had 'to compete with other companies who are privileged to employ Chinese labor, such as the great milling companies along the coast, and this company would, of course, be gravely handicapped in its competition with those institutions.' The bill passed without an anti-Chinese clause.[38]

In the spring of 1890, a number of workers at Royal City Planing Mills on False Creek struck over the quality of food served in the company's boarding house, and in the fall of 1891 Japanese sawmill workers struck for higher wages: much to the consternation of the Vancouver Trades and Labour Council 'their places were filled by white men at still less wages than what were struck for.' There were about sixty to seventy Japanese sawmill workers in Vancouver at this time. The Michigan Lumber Company reduced its Chinese workforce from between thirty and forty down to one or two in the spring of 1892 'on account of Mongolian labor not being of pecuniary benefit.' The real reason behind the lay-off is not known. The sluggish lumber market in 1894 prompted the Hastings mill to reduce the wages of its employees, but a similar cut at Royal City Planing Mills in New Westminster was met by a strike on 1 December 1894. The workers were willing to negotiate, but the company was intransigent, eventually reopening the sawmill by persuading a number of skilled, old hands to return to work at the reduced wage rates, and bringing in scabs to round out the needed workforce.[39]

Workers in the sash-and-door manufactories also engaged in job action after 1886, but they did not make common cause with the sawmill workers. Although they struggled for wages and shorter hours of work, a major aspect of their action was to retain their association with the carpenters engaged in the building trade. They wanted to be treated with the same respect as the carpenters, and did not want to be treated as lesser factory hands, who merely operated machines in woodworking plants. Employers, for their part, were keen to dissociate the manufactory workers from the carpenters and impose lower wages and longer hours of work.

On 4 July 1889, Vancouver carpenters went on strike to win the nine-hour workday, and workers in a number of sash-and-door manufactories joined in the job action. Within a day, some thirty building contractors agreed to the union's terms, but four contractors and two sash-and-door shops (Royal City Planing Mills Vancouver and Cassady & Company) refused to submit. The manufactory employers argued that shutting down their plants and machinery an hour earlier each day was too costly, and so the workers responded by suggesting that they take their hours off on Saturday, giving them Saturday afternoon free. Even this arrangement showed that the building trades and the workers in the manufactories operated in different environments. On the evening of 10 July, the employers and the union settled. The carpenters in the building trade achieved the nine-hour workday, and the manufactory carpenters were allowed to knock off at 3 o'clock on Saturdays.

On 13 July 1889, some seventeen or eighteen factory carpenters at the Vancouver Sash and Planing Mill struck to achieve the new terms, and the dispute was settled in favour of the workers on 17 July. The movement spread to New Westminster. On 2 April 1890, some thirty factory workers at Royal City Planing Mills went on strike. At the time carpenters in the construction industry were already working a nine-hour day. The sawmill, which ran on the ten-hour system, continued to operate, but Royal City Planing Mills capitulated on 4 April 1890.[40] Workers in the sawmill sash-and-door shops struggled to retain their designation as carpenters, but it was clear that the factory setting made them different from carpenters in the building trades, and that employers were keen to transform their job description. In the depression of the 1890s, there was no job action.

If there was conflict between millworkers and their employers, there was also evidence of cooperation, notable in the annual picnic organized by workers and management. In July 1890 Brunette Saw Mills allowed its employees to use the company steamer *Adelaide* for an excursion to Pitt Lake to celebrate Dominion Day. Later in the month, Royal City Planing Mills New Westminster chartered a steamer to take employees and their families down the Fraser River to Lulu Island for a picnic. About 500 people attended the event, and the steamer made two trips to transport the crowd. In the summer of 1891, the Royal City Planing Mills picnic was held at Stave River. The 1894 outing organized by British Columbia Mills Timber & Trading Company for its Vancouver workers was a trip to Chemainus on the east coast of Vancouver Island. The *City of Nanaimo* transported the assemblage across the strait, flying the house flag of the company, a white circular saw centred in a white field. The Vancouver city band was present, playing popular and classical airs during the journey. The company owner, John Hendry, and the Hastings manager, R.H. Alexander, were present, and upon arrival at Chemainus the excursionists ate a sumptuous lunch 'provided by a general cooperation on the part of the committee and the ladies, sweethearts and wives of the excurionists.' The crowd then moved to the recreation grounds, where games and sports were played. In the baseball game the Hastings mill team defeated Royal City Planing Mills Vancouver by a score of 18 to 7. The return trip began at six o'clock, with all looking 'forward to the sawdust men's outing next year.'[41]

Despite the racial and skill divisions, on rare occasions after 1900 their identity as workers won out and solidarity was achieved.[42] This was a difficult feat even among skilled white workers. Steam engineers in the mills, who were accredited by the government, were keen to increase their wages, assert their status as skilled workers, and lessen their hours of work. They blamed their inability to gain a shorter workday on the Asians, who they thought were too willing to keep the mills running for ten hours per day. The steam engineers had to be on the job even longer than the production workers, as they had to get up the steam before the mill was running and shut it down and take care of repairs after the end of the production day.

Nor were the steam engineers sympathetic towards their fellow white skilled workers such as the sawyers and the filers. They could not understand why the sawyers and filers were paid more than they were, since the sawyers and filers had not passed an examination and did not even have a formal certificate from the government recognizing their qualifications. William Ditcham, the Grand Chief of the British Columbia Association of Stationary Engineers, an organization that was against striking to win its aims, derisively offered that sawyers had no special qualifications 'except shoving through the lumber so as to cut the stuff good.'[43]

In December 1900, about 75 white and 100 Japanese workers at Royal City Planing Mills, Vancouver, went on strike for three days, and as a result postponed a wage reduction for four months.[44] In March 1901, 25 shingle sawyers in New Westminster struck against their employer for an increase in the piece rate from ten to fifteen cents per thousand shingles.[45] A two-day strike beginning 15 January 1903 at the Brunette Saw Mill ended in victory for the workers. The company wanted to reduce the number of foremen from one per every 30 men to one per every 100 men. The Japanese workers objected, and about 100 Japanese workers and 50 white workers went on strike.[46] In May 1903, at the same mill, some 20 Japanese workers struck for a day because they were being overcharged for supplies by their boss.[47]

Mill owners were capable of forcefully dealing with strikes. In 1903 workers in the sash-and-door manufactories, who were working ten-hour days, announced that as of 1 June they would work only fifty hours per week, made up of nine-hour days from Monday to Friday, and a five-hour day on Saturday, giving them Saturday afternoon off. Carpenters in the building trade had just secured the eight-hour work-day after a strike. The British Columbia Lumber and Shingle Manufacturers' Association intervened in this strike to thwart union growth. The mills persuaded the building contractors not to hire any strikers, and they reinforced their demand on the building contractors by refusing to sell lumber to any contractor that employed union carpenters. They were just as adamant in dealing with their own employees. The affected employers, Royal City Planing Mills in Vancouver and New Westminster, and Vancouver Sash-and-door, Robertson & Hackett, and Heaps & Co., all of Vancouver, refused to lower the hours of work. A strike ensued involving about 90 men in Vancouver 'and 37 white men and 20 Japanese and Chinese' in New Westminster. These were machine workers and bench hands. Workers finally called the strike off on 14 July, returning to work under the companies' terms.[48]

In early 1905, Japanese workers employed in a New Westminster sawmill were notified that their wages were to be reduced by $0.20 per day. They told management that they were prepared to strike, and the mill manager revoked the order. Japanese workers at the North Pacific Lumber Company mill in Barnet also struck in 1905, once to protest the dismissal of a worker and again because of a dispute with a contract boss.[49] In the spring of 1906 some sixty Chinese millworkers in Victoria demanded an increase in wages, from $1.25 to $1.50 per day, and a compromise was reached with the men accepting $1.35 per day. On 1 October 1907, mill owners and stevedore companies locked out eighty-five lumber handlers at two mills in Vancouver and New Westminster, because the workers refused to work ten hours per day instead of the current nine. Workers returned to the job on 23 October under their preferred hours of work. On 22 February 1909, about 100 East Indian workers at Fraser River Sawmills struck because they were dissatisfied with the conduct of an East Indian foreman/contractor. The foreman was not giving them their wages. Two hundred non–East Indian workers remained on the job.[50] Conflict between white and Asian workers was evident at a small shingle mill on the Pitt River in 1909, when ten white and fifteen Japanese workers 'used clubs, crowbars, and similar weapons to inflict painful, but not serious injuries on each other.'[51]

If job action in the sawmills was limited in the early years of the twentieth century, a time of general labour unrest in British Columbia, union organizations, too, were weak and fleeting. White sawyers and filers participated in the ineffective Pacific Coast Shingle Weavers Union in 1900 that wanted to prevent the hiring of Chinese workers as sawyers. The American Federation of Labor chartered a woodworkers' union, which attempted to organize New Westminster shingle and sawmill workers in 1903, but it disappeared within a few years. In 1907 there was evidence of an organizing drive among Asian workers: a white labour activist visited the mills and was surprised to find that Chinese and Japanese workers in almost all of the sawmills on Burrard Inlet and False Creek were already organized. The Asians wanted the whites to organize so that a general demand for higher wages could be put forward.[52] A lumber handlers' union was dissolved in October 1909. In 1911 workers at the large Canadian Western Lumber Company, Fraser Mills, organized a union, with New Westminster alderman John H. Campbell as president. In 1913 the International Union of Shingle Weavers, Sawmill Workers and Woodsmen, a U.S. union, organized three locals in Canada, two of which were in British Columbia. The Canadian locals contained only fifty members.[53]

With the onset of the economic downturn, lay-offs, wage cuts, and shorter hours became the order of the day. By August 1913 almost all the shingle mills in the province were closed due to market conditions, and sawmills in Vancouver reduced working hours from fifty to forty per week. In September, the Canadian Western Lumber Company discontinued its night shift and reduced wages. There was a further wage cut before the end of the year. In the late summer of 1914, Canadian Western Lumber discharged nearly 200 men, cut all wages by 10 to 20 per cent, and reduced the price of shingles to 13 cents, causing almost all the shingle weavers to quit. Vancouver Island mills were also closing to cope with the economic downturn.[54] Only with the revival of trade after 1915 was there a new round of union organization and strike activity in coastal mills.[55]

Race, ethnicity, and skill defined the mill workforce between 1858 and 1913. Prior to 1885, ethnicity and skill were intertwined as white Canadian and British workers controlled the best-paid positions. After 1885 the split was between Asian workers and white workers, with the white workers benefiting from the presence of the poorly paid Asians. Then there were the fluctuating markets which imposed a stern discipline on workers. Despite some attempts to overcome divisions, success was meagre in the years before the First World War.

Conclusion

In the late 1870s, Thomas George Askew's sawmill operation on the east coast of Vancouver Island was in serious financial trouble. A poor migrant from England, Askew began lumbering at Chemainus in the early 1860s, after a stint in the Cariboo goldfields.[1] Timber supply was not the problem – his mill was situated in a region that had remarkable forests. But Askew was unsuccessful. From the outset, his operation was poorly capitalized, and in 1864 he began receiving loans from Charles Edward Searle, the Anglican curate in his home village in England. But even with the loans, Askew was unable to establish market links. Searle wrote in 1867: ''Tis a pity that you cannot find a market for what I have heard from other sources is superior to anything of the sort in the world. But time is always required to develope the merits of a new market; for men are generally ignorant of foreign countries ... and further are very suspicious and fancy some cute trick is going to be played on them.'[2] In 1867 Askew arranged a lumber consignments through Gilbert Malcolm Sproat, using Searle as a reference, but the fate of this arrangement is unclear.

Searle also tried to use his family connections to help Askew. Searle's brother was associated with a London-based merchant house, Goodliffe, Smart & Searle, which had offices in Cape Town, at the southern tip of Africa. However, despite discussions, no contracts were signed, and Searle wrote that his brother's partners 'are rather safe men and discourage speculations.'[3] In 1872 Askew appealed to Searle to personally secure him spar contracts in England, but Searle, who had an ill mother and who had no business connections – he saw his brother rarely – was not helpful. At about the same time, Askew lost money on a spar contract, incurring debts to R.P. Rithet, the Victoria merchant, and

Francis A. Horne. By 1873 Askew was in serious financial difficulties, and his money problems were taking a toll on his health. He felt that the construction of a railway between Victoria and Nanaimo was now his only hope. Rumours abounded that a railway line was to be constructed between Victoria and Nanaimo, passing through the Chemainus region. Askew overextended himself financially, buying land in hopes of cashing in later. The railway did not come in time. On 27 January 1879, Askew was forced to put up his property as collateral for a mortgage of $3,500, and in September he sold out.[4] Askew was bankrupt and died in 1880.

Askew's failure illustrates a number of features of the early difficult years of lumbering in coastal British Columbia. With a small local market and limited access to the United States, connections to global markets were important. Askew's inability to link with a merchant such as Rithet, Andrew Welch, or Searle's brother, all with international correspondents, was a serious failing. The Chemainus region alone was apparently too small, and for unknown reasons Askew was unable to penetrate the nearby Victoria market. He was too far from New Westminster, which in the 1870s was a good environment for the expansion and success of small, personal firms. Moreover, Askew was unable to diversify into other operations such as farming or mining to sustain lumbering in the slow years. Perhaps the construction of the Esquimalt & Nanaimo Railway in the 1880s would have been his economic salvation. But it came too late.

Other small operators in the region remained in business in these lean years, and some even expanded as the 1870s and 1880s progressed. As the local market assumed greater importance, the merchant-controlled export firms, with their international connections, relinquished the most prominent role in the lumber economy to entrepreneurial companies that were owned and managed by people living in the region. The rise of the Canadian prairies as a significant market for lumber products after 1898 further enhanced the role of locally based producers. Corporate conglomerates, subsidiaries of lumber companies based outside the region, financial capitalists, and small personal enterprises operated lumbering firms, but firms at the entrepreneurial stage were the most dynamic in the sector. All firms operated in a competitive, decentralized economic environment.

For producers, even boom times were a mixed blessing. The promise of profit brought new entrants into the industry, leading to overproduction and small profit margins. Larger firms in the industry could not stabilize the situation through market, resource, or technological dominance. Cooperation in trade associations was a tactic that began in the mid-1880s, but it, too, brought little success. Financial insecurity remained an important aspect of the lumber business.

Governments, beginning in colonial times, encouraged the development of the lumber industry. Because the fear of speculation ran deep and the commitment to industrial activity was high, timber in the early years was allocated only to companies that operated sawmills. Thus, until the early twentieth century, sawmill companies controlled logging as well as milling. Governments tinkered with timber-allocation procedures in the nineteenth century, but because the growth of the industry was slow, these measures were harmless, and the basic strategy established in the 1860s remained in place.

In the first years of the twentieth century, changes in the relationship between logging companies and milling firms, as well as in the continental lumber economy, led to provincial timber legislation that threatened to further destabilize the industry. In an attempt to meet the needs of the new independent logging companies that were not tied to particular mills, operating in a quasi-open log market, the government changed the terms of the timber licences. The timber licence had been introduced in the 1880s but was of little significance until the boom years after 1898. Public concern, and more importantly, problems for industry participants, forced the government to rethink its new timber licensing regulations. As usual, in the end industry interests were satisfied. The scientific conservation movement, which informed the political discourse of the early twentieth century, influenced the language of government but did not challenge the capital structure of the industry. Unlike the nineteenth-century populist critique, which located environmental concerns in a political and economic analysis of an expanding industrial society, the scientific conservationists and the large capitalists in the industry were closely aligned by 1913.

A blinkered fascination with markets, capitalization, and government policies ignores the degree of industrialization that occurred in the coastal lumber industry in the half-century before 1913. Steam power, new machines and techniques, a division of labour, large units of production, factory settings, efficient coordination of various phases of the production process, and management strategies to control the labour process were all evident by 1913 in both the woods and the mills. Moreover, loggers and millworkers, members of the working class, were beginning to articulate their place in the industrial capitalist order.

Because they were not at the forefront of union organization and political action in British Columbia in the years before 1913, lumber-industry workers have largely been ignored. In the woods, loggers were overwhelmingly young, single, and white, dispersed in isolated camps throughout the coastal region. Work was dangerous and jobs were distinguished by skill levels. There was also a growing debate about who loggers were, a discourse that involved defining the meaning of the loggers' way of life: Was he a carefree individual, a man apart from other workers, or was he an industrial labourer who needed the benefits of organization and unions in a society dominated by corporations? Employers and union organizers engaged in the debate, which included probing the meaning of masculinity and domesticity. In the mills, race and skill were intertwined after the 1880s to give the coastal lumber industry a particular cast, which undermined solidarity. The attempts to organize lumber workers accelerated after 1913, but it was not until 1937, with the formation of the British Columbia district of the International Woodworkers of America, that divisions were overcome organizationally. Both millworkers and woods workers were part of this union.[5]

Market relations and the process of industrialization are both important aspects of the history of the lumber industry. The market orientation of staples theory offers important avenues for investigation. The legislative demands and transportation requirements of resource companies; the role of external capital, markets, and labour; the geographical flow of profits; and the location of research and development facilities raise important questions about economies based on staple exploitation. The larger question of the future of these economies remains pertinent, especially in a province like British Columbia, where resource industries continue to be important in many communities. Similarly, theories of industrialization inform understandings of the history of corporate developments, the dynamic of class relations, and the changing nature of the production process in a staples economy. In recent years, the depletion of the resource base, environmental and ecological concerns, as well as new technologies, business organizational structures, global market conditions, and government strategies have transformed the debate about resource industries. Yet, in this post-Fordist era, perhaps more than ever, we must still attempt to understand the complex processes that got us here.

Notes

Introduction

1 For overviews of the history of British Columbia, see Margaret A. Ormsby, *British Columbia: A History*; Jean Barman, *The West beyond the West: A History of British Columbia*; Hugh J.M. Johnston, ed., *The Pacific Province: A History of British Columbia.*
2 Joseph Collins Lawrence, 'Markets and Capital: A History of the Lumber Industry of British Columbia (1778–1952).'
3 G.W. Taylor, *Timber: History of the Forest Industry in B.C.*
4 Ken Drushka, *HR: A Biography of H.R. MacMillan*; Donald MacKay, *Empire of Wood: The MacMillan Bloedel Story*; Sue Baptie, ed., *First Growth: The Story of British Columbia Forest Products Limited*; Gordon Gibson, *Bull of the Woods: The Gordon Gibson Story*; E.G. Perrault, *Wood & Water: The Story of Seaboard Lumber and Shipping.*
5 Ken Bernsohn, *Cutting Up the North: The History of the Forest Industry in the Northern Interior*; Peter R. Gillis and Thomas R. Roach, *Lost Initiatives: Canada's Forest Industries, Forest Policy and Forest Conservation*; Stephen Gray, 'The Government's Timber Business: Forest Policy and Administration in British Columbia, 1912–1928'; Jeremy Wilson, 'Forest Conservation in British Columbia, 1935–85: Reflections on a Barren Political Debate.'
6 Myrtle Bergren, *Tough Timber: The Loggers of B.C. – Their Story*; Jerry Lembcke and William M. Tattam, *One Union in Wood: A Political History of the International Woodworkers of America*; Susanne Clausen, 'The Plywood Girls: Women and Gender Ideology at the Port Alberni Plywood Plant, 1942–1992'; Gordon Hak, 'British Columbia Loggers and the Lumber Workers Industrial Union, 1919–1922' and 'Red Wages: Communists and the 1934 Vancouver Island Loggers Strike.'

7 There are exceptions, of course. See Robert A.J. McDonald, 'Lumber Society on the Industrial Frontier: Burrard Inlet, 1863–1886'; Richard Rajala, 'The Forest as Factory: Technological Change and Worker Control in the West Coast Logging Industry, 1880–1930'; W. Kaye Lamb, 'Early Lumbering on Vancouver Island, Part I: 1844–1855' and 'Early Lumbering on Vancouver Island, Part II: 1855–1866.' Much interesting information, as well as many perceptive observations, on the lumber industry in Vancouver are found throughout Robert A.J. McDonald, *Making Vancouver: Class, Status, and Social Boundaries, 1863–1913*. Provincial government timber policy is touched on in Robert E. Cail, *Land, Man, and the Law: The Disposal of Crown Lands in British Columbia, 1871–1913*.

8 Like the literature on British Columbia, histories of forestry in the U.S. Pacific Northwest tend to focus on labour, business, or government. The work of William G. Robbins is more analytical than most, integrating a number of themes in a political-economy approach: see *Lumberjacks and Legislators: Political Economy of the U.S. Lumber Industry, 1890–1941*, and *Hard Times in Paradise: Coos Bay, Oregon, 1850–1896*. Some studies treat both the British Columbia and the U.S. Pacific Northwest. See Lembcke and Tattam, *One Union in Wood*, and Richard A. Rajala, *Clearcutting the Pacific Rain Forest: Production, Science, and Regulation*.

9 For a study of Kootenay mining in the context of global capitalism, see Jeremy Mouat, *Roaring Days: Rossland's Mines and the History of British Columbia*. A First Nations perspective on colonization is provided by Keith Thor Carlson, *You Are Asked to Witness: The Stó:lō in Canada's Pacific Coast History*. A broader perspective on colonization in the Far West is found in Cole Harris, *The Resettlement of British Columbia: Essays on Colonialism and Geographical Change*.

10 For contemporary descriptions of the coastal region, see R.E. Gosnell, *The Year Book of British Columbia and Manual of Provincial Information*, 231–5; *The Canada Lumberman, Wood-Workers', Manufacturers', and Millers' Gazette* [hereinafter *The Canada Lumberman*] 20/10 (October 1899): 25; *Canada Lumberman and Woodworker* 27/2 (February 1907): 43; H.N. Whitford and Roland D. Craig, *Forests of British Columbia*.

11 *The Canada Lumberman*, 21/5 (May 1901): 12.

12 There was large-scale lumbering in the Alberni Valley in 1912. A railway, completed in December 1911, linked the mill to prairie markets, but the area was still separate from the coastal industry that hugged the shores of the inside passage: Gordon Hak, 'On the Fringes: Capital and Labour in the Forest Economies of the Port Alberni and Prince George Districts, British Columbia, 1910–1939,' 38–42.

13 By 1913 lumber economies had emerged in the northern interior, and the Kootenay or mountain region of the province. Smaller trees, lack of ocean access, and harsher winters dictated different patterns of development. North of the coastal region under discussion here, there was also interest in lumbering in the Queen Charlotte Islands, or Haida Gwaii, and the northwest region of the province. In 1909 pulp-and-paper production began in the coastal region, but it plays only a peripheral role in this study.

14 The Forest Act of 1912, a serious economic depression, the opening of the Panama Canal and the outbreak of the First World War in 1914, and the introduction of high-lead logging also make 1913 a meaningful end point for this study.

15 A helpful introduction to Innis is Daniel Drache, ed., *Staples, Markets, and Cultural Change: Selected Essays / Harold Innis.*

16 A.R.M. Lower, 'The Trade in Square Timber,' 28.

17 W.A. Carrothers, 'Forest Industries of British Columbia.'

18 For an outline of staple theory, see M.H. Watkins, 'A Staple Theory of Economic Growth.' For an optimistic view of staples-led growth and development, see W.A. Mackintosh's 1923 article 'Economic Factors in Canadian History.'

19 For useful criticisms of the staples perspective, see Gregory S. Kealey, *Toronto Workers Respond to Industrial Capitalism, 1867–1892,* 3–34; Wallace Clement, *Hardrock Mining: Industrial Relations and Technological Changes at Inco,* 15–26; Graham D. Taylor and Peter A. Baskerville, *A Concise History of Business in Canada,* xi–xvi.

20 See Thomas I. Gunton, *Resources, Regional Development and Provincial Policy: A Case Study of British Columbia,* 4–17; Patricia Marchak, *Green Gold: The Forest Industry in British Columbia,* 1–27.

21 Trevor J. Barnes and Roger Hayter, eds., *Troubles in the Rainforest: British Columbia's Forest Economy in Transition.*

22 Patricia Marchak, 'A Changing Global Context for British Columbia's Forest Industry,' 149.

23 For a theoretical perspective that integrates both exchange and production in defining the historical development of capitalism, see Dale Tomich, 'World of Capital/Worlds of Labor: A Global Perspective.'

24 'Woods Work.'

1: Markets

1 W.A. Carrothers, 'Forest Industries of British Columbia,' 263n.

2 Nelson Courlandt Brown, *The American Lumber Industry, Embracing the*

Principal Features of the Resources, Production, Distribution, and Utilization of Lumber in the United States, 229.

3 Ibid.

4 James E. Hendrickson, ed., *Journals of the Colonial Legislatures of the Colonies of Vancouver Island and British Columbia, 1851–1871*, vol. 2: 5–6.

5 W. Kaye Lamb, 'Early Lumbering on Vancouver Island, Part I: 1844–1855,' 47.

6 Thomas R. Cox, *Mills and Markets: A History of the Pacific Coast Lumber Industry to 1900*, 301.

7 Joanne Reitano, *The Tariff Question in the Gilded Age: The Great Debate of 1888.*

8 *The Truth*, 22 August 1890; *Daily Herald*, 3 January 1888.

9 *Northwestern Lumberman*, reprinted in *Vancouver Daily World*, 19 January 1892.

10 Reprinted in *Vancouver Daily World*, 30 December 1893.

11 *Vancouver Daily World*, 19 August 1895.

12 *The Daily Columbian*, 14 March, 15 March, 11 July, 10 October 1895.

13 Lamb, 'Part II,' 48.

14 Brown, *The American Lumber Industry*, 229; Robert Griffin, 'The Shingle Sawing Machine in British Columbia, 1901 to 1925'; Testimony of Thomas F. Paterson, Transcript, Timber and Forestry Commission, 1909, British Columbia Archives and Records Service [hereinafter BCARS] GR 271, v. 2, f. 2, p. 290; Memorandum, n.d., BCARS, GR 271, v. 1, f. 10.

15 *The British Columbian*, 11 April 1865.

16 *The British Columbian*, 17 October 1866.

17 *The British Columbian*, 19 November, 26 November 1864.

18 R.P. Rithet to Mr Sproat, 6 April 1869, BCARS, 'Letterbook Correspondence, 15 October 1868 to 9 September 1871,' Rithet Family Papers, Add MSS 504, vol. 2.

19 'J.R. Stewart,' BCARS, GR 1372, B 1366, f. 1655; Rithet to Welch, 24 August 1871, BCARS, Add MSS 504, vol. 2.

20 Mr Owen, British Consul at Port Leonka, Fiji, to Earl Russell, Foreign Minister, 28 April 1864, printed in *The British Columbian*, 11 January 1865.

21 *The British Columbian*, 8 June 1867.

22 *The Mainland Guardian*, New Westminster, 29 January, 21 May, 13 July 1870.

23 *Dominion Pacific Herald*, 5 April 1879.

24 *The Daily British Columbian*, 20 June, 10 November 1887; 10 May 1888.

25 *The Daily British Columbian*, 21 February 1888.

26 *The Mainland Guardian*, 6 June 1888.

27 *The Mainland Guardian*, 25 February, 6 June 1888; *The Daily British Columbian*, 2 March, 18 June 1888.
28 *Canada Lumberman and Woodworker* 27/2 (February 1907): 41.
29 *The Labour Gazette* 3 (September 1902): 161.
30 *The Canada Lumberman Wood-Workers', Manufacturers', and Millers' Gazette* [hereinafter *The Canada Lumberman*] 24/2 (September 1904): 25; *The Labour Gazette*, 6 (June 1906): 1345. The role of exports in the British Columbia economy as a whole during the boom of 1900 to 1914 was not that great: 'In British Columbia ... exports definitely did not lead the van: at the peak of the boom, in 1913, the value of British Columbia's exports was up only 25 per cent over 1901': Allen Seager, 'The Resource Economy, 1871–1921,' 234.
31 Joseph Collins Lawrence, 'Markets and Capital: A History of the Lumber Industry of British Columbia (1778–1952),' 9–11.
32 Thomas Donahue to Governor Douglas, 2 February 1860, BCARS, GR 1372, B 1325, f. 479; *The British Columbian*, New Westminster, 23 December 1863.
33 Lamb, 'Part I,' 31–42; Richard Somerset Mackie, 'Colonial Land, Indian Labour and Company Capital: The Economy of Vancouver Island, 1849–1858,' 141–50; 203–9.
34 Wm. P. Sayward to Hon. Mr Birch, Colonial Secretary, 21 March 1867, BCARS, Colonial Correspondence, GR 1372, B 1364, f. 1570; Mackie, 'Colonial Land,' 221–2.
35 Jean Barman, *The West beyond the West: A History of British Columbia*, 363 and 371.
36 Keith Ralston, 'Patterns of Trade and Investment on the Pacific Coast, 1867–1892: The Case of the British Columbia Canning Industry,' 300; Cole Harris, *The Resettlement of British Columbia: Essays on Colonialism and Geographical Change*, 92–6.
37 *Port Moody Gazette*, 22 December 1883.
38 *Port Moody Gazette*, 22 December 1883; 15 March, 9 August 1884.
39 *The British Columbian*, 15 March 1882.
40 *The British Columbian*, 22 July, 20 September 1882; 29 March 1884.
41 *The British Columbian*, 16 August 1882; *The Mainland Guardian*, 22 December 1883.
42 Jean Barman, *The West beyond the West*, 363, 371, and 374.
43 British Columbia, *Sessional Papers, 1891* (Victoria: Queen's Printer, 1891), 280–280A.
44 Barman, *The West Beyond the West*, 374–5.
45 *The Daily British Columbian*, New Westminster, 21 August 1888.

46 *The Daily British Columbian*, 17 September, 30 September 1886; 3 January, 11 September 1888; *The Mainland Guardian*, New Westminster, 8 February, 22 February, 29 February, 3 March, 10 March, 24 March, 28 March, 21 April, 2 June, 9 June, 16 June 1888.

47 D.N. Sprague, *Post-Confederation Canada: The Structure of Canadian History since Confederation*, 370.

48 Ibid.

49 Timber and Forestry Commission, 1909, Papers, BCARS, GR 271, v. 1, f. 8.

50 Lawrence, 'Markets and Capital,' 96 and 99.

51 *The Canada Lumberman* 19/3 (March 1898): 11.

52 *The Canada Lumberman* 19/5 (May 1898): 11; 24/1 (January 1904): 12–13. The negative voters, from the Liberal–Labour caucus, believed that free trade benefited Canadian workers in general. Rather than supporting workers in the lumber industry, who would ostensibly benefit in the protected market with greater job security, they argued that all workers gained from free access because of lower prices. Allowing U.S. lumber producers to ship duty-free into Canada undermined what they perceived as the monopolistic tendencies among British Columbia millmen, keeping lumber prices low, to the advantage of local construction workers, private home builders, and other industries in the province that required lumber.

53 *The Canada Lumberman* 21/3 (March 1901): 8.

54 Minutes, British Columbia Branch of the Canadian Manufacturers' Association, Vancouver, 19 March 1903, University of British Columbia, Special Collections Division, Council of Forest Industries Papers, v. 31, f. 8.

55 *Canada Lumberman and Woodworker* 25/3 (March 1905): 23.

56 *Western Lumberman* 7/9 (September 1910): 21–2.

57 *Western Lumberman* 6/1 (January 1909): 17; 6/3 (March 1909): 11–12; 6/9 (September 1909): 13; Phyllis Veazey, 'John Hendry and the Vancouver, Westminster and Yukon Railway: "It Would Put Us on Easy Street,"' 59; William G. Robbins, *Lumberjacks and Legislators: Political Economy of the U.S. Lumber Industry, 1890–1941*, 65–7.

58 L.R. Andrews, Memorandum on Tariff Policies, Exhibit 220, Royal Commission on Forestry, 1944–5, Proceedings, BCARS, GR 520, v. 13, f. 11.

59 Memorandum, BCARS, Timber and Forestry Commission, 1909, GR 271, v. 1, f. 10. According to another account, in British Columbia as a whole in the years from 1901 to 1913, 63–75 per cent of lumber production was shipped east, 5–17 per cent was exported overseas, and the remainder was consumed in the province: R. Peter Gillis and Thomas R. Roach, *Lost Initiatives: Canada's Forest Industries, Forest Policy and Forest Conservation*, 134.

2: Mill Companies

1 See G.W. Taylor, *Timber: History of the Forest Industry in B.C.* Chapter titles include 'The Coming of the Easterners' and 'The American Invasion, 1900–1914.'

2 This sketch of the development of industrial capitalism draws on Craig Heron and Robert Storey, 'On the Job in Canada'; Craig Heron, 'The Second Industrial Revolution in Canada, 1890–1930'; Ian McKay, 'Capital and Labour in the Halifax Baking and Confectionary Industry during the Last Half of the Nineteenth Century'; Gregory S. Kealey, *Toronto Workers Respond to Industrial Capitalism, 1867–1892*, 2–34; Craig Heron, 'Factory Workers'; Alfred D. Chandler, Jr, *The Visible Hand: The Managerial Revolution in American Business*, 9–10. Robert A.J. McDonald uses Chandler's theory of firm growth to analyse the Vancouver business community in *Making Vancouver: Class, Status and Social Boundaries, 1863–1913*, 62–89.

3 Paul Phillips, 'The Underground Economy: The Mining Frontier to 1920'; Jeremy Mouat, *Roaring Days: Rossland's Mines and the History of British Columbia*; David J. Reid, 'Company Mergers in the Fraser River Canning Industry, 1885–1902'; Robert A.J. McDonald, 'Victoria, Vancouver, and the Economic Development of British Columbia, 1886–1914.'

4 *The Labour Gazette* 6 (June 1906): 1344–5.

5 'Diary of a Sawyer,' May to October 1902, British Columbia Archives and Records Service [hereinafter BCARS], Pidcock Family Papers, Add Mss 728, v. 1.

6 See J.M.S. Careless, 'The Lowe Brothers, 1852–70: A Study in Business Relations on the North Pacific Coast,' and 'The Business Community in the Early Development of Victoria, British Columbia.'

7 Dickson, DeWolf & Company to C.W. Wallace, 25 September 1866, BCARS, GR 1372, B 1325, f. 471a.

8 Edward Stamp to James Douglas, 21 December 1859, BCARS, GR 1372, B 1366, f. 1643.

9 Wm. Banfield to W.A.G. Young, 3 July 1860, BCARS, GR 1372, B 1306, f. 107.

10 W. Kaye Lamb, 'Early Lumbering on Vancouver Island, Part II: 1855–1866,' 100–1.

11 Wm. Banfield to Colonial Secretary, 10 August 1861, BCARS, GR 1372, B 1306, f. 107; R.C. Mayne, *Four Years in British Columbia and Vancouver Island: An Account of Their Forests, Rivers, Coasts, Gold Fields, and Resources for Colonisation*, 228.

12 Matthew Macfie, *Vancouver Island and British Columbia, Their History, Resources, and Prospects*, 135.

13 Edward Stamp to W.A.G. Young, 22 August 1861, BCARS, GR 1372, B 1366, f. 1643.

14 Anderson & Co. to J.D. Pemberton, 18 February 1863, BCARS, GR 1372, B 1300, f. 17.

15 Gilbert Malcolm Sproat to the Colonial Secretary, 1 November 1864, BCARS, GR 1372, B 1366, f. 1638.

16 Considering that the Alberni region became a major producer of lumber, and pulp and paper, in the twentieth century, the closure of the mill due to a shortage of timber seems peculiar. Not all historians have been convinced by this explanation. G.W. Taylor argues that the operation closed because the mill owners 'refused to put any money into it': Taylor, *Timber*, 22. Joseph Collins Lawrence links the closure of the operation to the end of the American Civil War: 'the bulk of the British trade in timbers again reverted to Southern ports and operations at Alberni dwindled and finally came to a standstill' (Lawrence, 'Markets and Capital: A History of the Lumber Industry of British Columbia [1778–1952],' 23). Thomas R. Cox is also sceptical of the timber-shortage thesis: 'The real problem appears not to have been a shortage of sawlogs but the tariff barriers that generally kept the firm's lumber out of San Francisco and left it without markets large and dependable enough to keep it running during slack times as well as flush' (Cox, *Mills and Markets: A History of the Pacific Coast Lumber Industry to 1900*, 134).

W. Kaye Lamb, who actually delved into the archival records, remains the most persuasive: 'Sproat's letters make it clear that it was closed simply because it did not yield a profit, and because the available supply of timber was exhausted.' He also notes that the supply of timber depends on the ability to transport logs profitably from the forest to the mill site: 'it must be remembered that in 1864 the only way to convey a log to the mill was to drag it with oxen or float it down a stream. Sproat's failure does not prove that he was incompetent; it simply reveals the extent to which the lumber industry of the present is the product of modern logging methods and machinery.' Greater investment would have increased the accessible timber supply, and even Sproat contemplated constructing a 'railway to convey logs for many miles,' but the costs were prohibitive. There is no evidence that the Civil War had any impact on the fate of the mill. The U.S. tariff was indeed a factor, but ultimately, unlike other mill operations on Vancouver Island and in British Columbia, Anderson & Co. were unwilling to run their facility at a loss: Lamb, 'Part II,' 14, 15, and 110; Sproat to Colonial Secretary, 1 November 1864, BCARS, GR 1372, B 1366, f. 1638.

17 Edward Stamp to W.A.G. Young, 18 June 1863; 22 September, BCARS, GR 1372, B 1366, f. 1643.

18 Edward Stamp to Arthur N. Birch, 17 May 1865, BCARS, GR 1372, B 1366, f. 1643.

19 F.W. Howay, 'Early Shipping Burrard Inlet: 1863–1870,' 8–11; Captain E. Stamp to Hon. A.N. Birch, 18 July 1865, University of British Columbia, Special Collections Division [hereinafter UBC], Hastings Saw Mill Correspondence, v. 1, f. 9; Edward Stamp to A.N. Birch, 31 January 1867, UBC, Hastings Saw Mill Correspondence, v. 1, f. 36.

20 J.C. Nicholson to the Chief Commissioner of Lands and Works, 21 October 1870, BCARS, GR 1372, B 1350, f. 1251.

21 Anderson & Co. at Alberni was also a British trading firm. At their Victoria office they represented the Royal Insurance Company and sold British goods such as beer, as well as saws, files, chisels, and steel produced by a Sheffield firm. In London they were connected to Anderson, Thomson & Company: Invoice, Anderson & Co., 12 November 1862, BCARS, GR 1372, B 1300, f. 17. After the sawmill closed, the firm continued as importer-exporters in Victoria, at times taking on lumber consignments.

22 *The Mainland Guardian*, 29 May, 27 November 1872; 4 April 1874.

23 *The British Columbian*, 5 July 1884; Memorandum of Association, 29 September 1887, UBC, Hastings Saw Mill Correspondence, v. 1, f. 60.

24 George Waite Stirling Brooks, 'Edgar Crow Baker: An Entrepreneur in Early British Columbia,' 196; *The Mainland Guardian*, 24 September 1884.

25 Brooks, 'Edgar Crow Baker,' 4, 77, 185–6.

26 Ibid., 192.

27 Involved with the syndicate were the following Victorians: C.T. Dupont, D.R. Harris, Robert E. Jackson, E.G. Prior, George A. Keefer, F.S. Barnard, Carl Strouss, Israel Wood Powell, William C. Ward, and perhaps W.P. Sayward. David Oppenheimer, Edgar Baker, A.W. Ross, P.C. Dunlevy, R.H. Alexander, and Alex Cambie were also associated with the group: *Port Moody Gazette*, 16 May 1885; Memorandum of Association, 29 September 1887, UBC, Hastings Saw Mill Correspondence, v. 1, f. 58; Brooks, 'Edgar Crow Baker,' 192–3.

28 *The British Columbian*, 2 August 1884.

29 *The British Columbian*, 22 October 1884.

30 *The Daily British Columbian*, 3 January, 9 July, 23 July 1887; 13 September 1889; *The Daily Herald*, Vancouver, 25 July, 8 August, 15 September 1887; *Vancouver Daily World*, 12 September 1889; 8 March 1890; Memorandum of Association, 29 September 1887, UBC, Hastings Saw Mill Correspondence, v. 1, f. 58; v. 1, f. 60.

31 *The British Columbian*, 15 August 1863; Howay, 'Early Shipping in Burrard Inlet,' 4–5.
32 Burrard Inlet Mill: Statement of Affairs [1863?], BCARS, GR 1372, B 1332, f. 665; *Mainland Guardian*, New Westminster, 20 December 1871.
33 *The British Columbian*, 2 December, 19 December 1863; 2 November, 19 November, 26 November 1864. In 1867 Moody & Company controlled this New Westminster mill: Arthur Harvey, *A Statistical Account of British Columbia*, 16.
34 *The British Columbian*, 2 December 1863. Ireland sold his interest in the company to Moody and Van Bramer in October 1865: ibid., 7 October 1865.
35 *The British Columbian*, 8 February, 23 March, 6 July 1865.
36 Howay, 'Early Shipping in Burrard Inlet,' 7.
37 Lawrence, 'Markets and Capital,' 25–6; *The British Columbian*, 15 April, 17 June, 23 September 1868; 2 July 1870.
38 'J.R. Stewart,' BCARS, GR 1372, B 1366, f. 1655; Rithet to Welch, 24 August 1871, BCARS, Add MSS 504, vol. 2.
39 R.P. Rithet to Raymur, 7 August 1871; R.P. Rithet to Mr Welch, 22 August 1871; R.P. Rithet to Clapperton, 28 August 1871, BCARS, Add MSS 504, vol. 2.
40 R.P. Rithet to Mr Welch, 22 August 1871, BCARS, Add MSS 504, vol. 2. The details on Rithet and the sale of Stewarts are from the letters in this volume as well.
41 G.W. Taylor, *Timber: History of the Forest Industry of B.C.*, 31.
42 *The British Columbian*, 21 June, 24 June 1882; 26 July 1889; Rithet to Henderson, 29 December 1884, BCARS, Add MSS 75, v. 1.
43 *The Daily British Columbian*, 9 March 1888; 14 June 1895; *Vancouver Daily World*, 5 June, 3 July 1891; *The British Columbia Commercial Journal*, 7 July 1891; Andrew Welch Death Certificate, BCARS, Rithet Family Papers, Add MSS 504, v. 4, f. 6.
44 *The Canada Lumberman, Wood-Workers', Manufacturers', and Millers' Gazette* [hereinafter *The Canada Lumberman*] 21/2 (February 1902): 13; 21/4 (April 1902): 15; 25/6 (June 1905): 13–14.
45 *The Mainland Guardian*, 13 July 1878; *Dominion Pacific Herald*, 5 May, 16 June 1880; *The Canada Lumberman* 24/7 (July 1904): 11 and 15; F.W. Howay, *British Columbia from Earliest Times to the Present: Biographical*, vol. 4, 720–5.
46 *The British Columbian*, 16 August, 1882; *The Mainland Guardian*, 22 December 1883.
47 *The British Columbian*, 4 August, 8 August, 22 September, 3 November 1883.
48 Dominion Saw Mill Co., Register of Shares, BCARS, Dominion Sawmill Co. Papers, Add MSS 75, v. 2.

49 *The Mainland Guardian*, 29 March 1884.
50 BCARS, Add MSS 75, v. 1.
51 Dominion Saw Mill Co., Register of Shareholders, BCARS, Add MSS 75, v. 3; *The Mainland Guardian*, 1 October, 15 October 1884; 16 May, 10 June, 13 June 1885; *The British Columbian*, 3 September 1884.
52 *Canada Lumberman and Woodworker* 25/6 (June 1905): 13–14.
53 Phyllis Veazey, 'John Hendry and the Vancouver Westminster and Yukon Railway: "It Would Put Us on Easy Street,"' 62n.
54 Ibid., 62.
55 George Ward DeBeck to Keary, 16 July 1931, BCARS, 'Correspondence re Brunette Sawmill'; *The Mainland Guardian*, New Westminster, 7 May, 8 October 1870; 23 September 1871; 23 October 1875; 28 November 1877; 16 March 1878; *Dominion Pacific Herald*, New Westminster, 24 April 1880; *The British Columbian*, New Westminster, 29 March 1884.
56 *The British Columbian*, 15 March 1882.
57 *The British Columbian*, 22 July, 20 September 1882; 29 March 1884.
58 *The Daily British Columbian*, 6 March 1886; 30 October 1889; 8 January 1891; 27 March 1894; 29 March, 9 February 1895; *The Truth*, New Westminster, 13 August 1890; *Vancouver Daily World*, 9 May 1894; Taylor, *Timber*, 45.
59 *The Mainland Guardian*, 3 November 1883; British Columbia, *Sessional Papers, 1891* (Victoria: Queen's Printer, 1891), 280–280B.
60 A. Haslam to W.S. Gore, Deputy Commissioner of Lands & Works, 19 January 1901, BCARS, Lands Branch Correspondence, GR 1440, B 2728, f. 354/01; f. 406/01; GR 1440, B 2755, f. 8446/04; *The Labour Gazette* 6 (October 1905): 410; (December 1905): 653; *Lumberman & Contractor* 2/12 (December 1905): 9; 3/12 (December 1906): 36.
61 Wm. P. Sayward to Hon. Mr Birch, Colonial Secretary, 21 March 1867, BCARS, Colonial Correspondence, GR 1372, B 1364, f. 1570; Richard Mackie, 'William Parsons Sayward,' 221–2.
62 B.W. Pearse to W.P. Sayward, 27 July 1869, BCARS, GR 1372, B 1340, f. 953.
63 *The Daily British Columbian*, 3 January, 9 July, 23 July 1887; 13 September 1889; *The Daily Herald*, 25 July, 8 August, 15 September 1887; *Vancouver Daily World*, 12 September 1889; 8 March 1890; *Daily Colonist*, Victoria, 19 September 1894; 5 April 1896; Robert Brian Griffin, 'The Shawnigan Lake Lumber Company, 1889–1943,' 14–21.
64 *The Canada Lumberman* 18/11 (November 1897): 15; *Canada Lumberman and Woodworker* 30/19 (1 October 1910): 28–9; Lawrence, 'Markets and Capital,' 80–1; Canadian Puget Sound Lumber Co., BCARS, K.R.Genn Collection, Add MSS 1950, v. 53, f. 1.
65 *The Daily British Columbian*, 10 December 1887; 28 May 1888; *The Mainland*

Guardian, 3 March, 6 October, 17 October 1888; *Vancouver Daily World*, 20 May 1890; *Morning Ledger*, New Westminster, 14 June 1891.

66 *Vancouver Daily World*, 13 July 1894; 23 October 1888; 14 October 1890; *The Daily British Columbian*, 6 January 1891; 30 May 1892; 11 December 1893; 12 May, 3 July 1894; 26 September 1895; *The Truth*, 19 September 1890; *Morning Ledger*, 13 June, 23 October 1891; James Maclaren to Chief Commissioner of Lands and Works, 3 January 1890, BCARS, Lands Branch Correspondence, GR 1440, B 2681, f. 175/90.

67 *Canada Lumberman and Woodworker* 25/2 (February 1905): 11; *Western Canada Lumberman* 5/4 (April 1908): 16; *Canada Lumberman and Woodworker* 30/21 (1 November 1910): 26–9; Taylor, *Timber*, 40–2.

68 Minutes, Shareholder Meeting, 4 April 1910, UBC, Humbird Family Papers, 9–1; Meeting of the Incorporators, Humbird Lumber Company, 13 December 1900, UBC, 9–4.

69 *Canada Lumberman and Woodworker* 25/9 (September 1905): 23–4; Lawrence, 'Markets and Capitals,' 89; Taylor, *Timber*, 93–4.

70 *Vancouver Daily World*, 13 July 1894; *The Canada Lumberman* 22/6 (June 1902): 7; *Canada Lumberman and Woodworker* 13/5 (1 March 1905): vi; 16/11 (November 1906): 28; 30/14 (15 July 1910): 34; *Lumberman & Contractor* 4/4 (April 1907): 22; Howay, *British Columbia from the Earliest Times to the Present*, 82–5; Wm. McNeill to Mr Hendry, 23 July 1910, Vancouver City Archives, McNeill Papers, Add MSS 46, v. 2, f. 8; Robert A.J. McDonald, 'Business Leaders in Early Vancouver, 1886–1914,' 138–9.

71 Gordon Hak, 'On the Fringes: Capital and Labour in the Forest Economies of the Port Alberni and Prince George Districts, British Columbia, 1910–1939,' 40–2.

72 *Vancouver Daily World*, 30 December 1893.

73 *The Daily Herald*, 15 March 1888; *Vancouver Daily World*, 13 July 1894; 19 March, 23 August, 25 August 1892; *The Daily British Columbian*, 31 May, 19 July 1889; 13 May 1893; 20 July 1894; Howay, *British Columbia from the Earliest Times to the Present*, 586–9 [William Lawrence Johnson].

74 *The Canada Lumberman* 22/2 (February 1902): 9.

75 *The Canada Lumberman*, 19/3 (March 1898): 8; 21/3 (March 1901): 8; James A. McNair to W.S. Gore, 9 June 1904, BCARS, GR 1440, B 2752, f. 4828/04.

76 'First Annual Report of the Inspector of Factories of the Province of British Columbia for the Year 1910,' British Columbia, *Sessional Papers, 1911* (Victoria: King's Printer, 1911), I9–I20.

77 This was not just a feature of the coastal lumber industry in British Colum-

bia, but was a North American phenomenon in lumbering: see Alfred D. Chandler Jr, *Scale and Scope: The Dynamics of Industrial Capitalism*, 45.

3: Business Strategies

1 Richard Mackie, 'Colonial Land, Indian Labour and Company Capital: The Economy of Vancouver Island, 1849–1858,' and *Trading beyond the Mountains: The British Fur Trade on the Pacific, 1793–1843.*
2 W. Kaye Lamb, 'Early Lumbering on Vancouver Island, Part I: 1844–1855'; Mackie, 'Colonial Land, Indian Labour and Company Capital, 141–50, 203–9.
3 Daniel T. Gallacher, 'John Muir,' 627.
4 Edward Stamp to W.A.G. Young, 22 August 1861; 12 February 1862, British Columbia Archives and Records Service [hereinafter BCARS], GR 1372, B 1366, f. 1643; Matthew Macfie, *Vancouver Island and British Columbia, Their History, Resources, and Prospects*, 135; Alexander Caulfield Anderson, *The Dominion at the West: A Brief Description of the Province of British Columbia, Its Climate and Resources*, Appendix B, i.
5 Anderson & Co. to W.A.G. Young, 22 January 1864; Anderson & Co. to the Colonial Secretary, 22 July 1864, BCARS, GR 1372, B 1300, f. 17; J.A. Raymur to W.A.G. Young, 12 March 1864, BCARS, GR 1372, B 1360, f. 1460.
6 Anderson & Co. to Henry Wakeford, 27 May 1865, BCARS, GR 1372, B 1300, f. 17.
7 Edward Stamp to A.N. Birch, 19 April 1865; Edward Stamp to Joseph W. Trutch, 23 July 1868, BCARS, GR 1372, B 1366, f. 1643.
8 *The British Columbian*, 7 October 1865.
9 *The British Columbian*, 5 July 1884.
10 *The British Columbian*, 26 August, 5 August 1882.
11 *The British Columbian*, 29 March 1884.
12 *The British Columbian*, 14 November 1883.
13 *The Mainland Guardian*, 30 August 1884.
14 Hastings Saw Mill Letterbook, 1870–4, Vancouver City Archives, Hastings Saw Mill, Add MSS 27.
15 *The Truth*, 13 August 1890; *Vancouver News*, 10 December 1886; 20 February 1887; *The Daily Herald*, 26 April, 9 June 1888; *Vancouver Daily World*, 13 July 1894; J.A. Sayward, Ledger 1892–4, BCARS, K.R. Genn Collection, Add MSS 1950, v. 165.
16 *Canada Lumberman and Woodworker* 30/19 (1 October 1910): 28; 25/6 (June 1905): 13–14; *The Timberman* 14/10 (August 1913): 31.

17 See R.T. Naylor, *The History of Canadian Business, 1867–1914:* Vol. 2: *Industrial Development*, 104–61.
18 *The British Columbian*, 30 January 1867.
19 James E. Hendrickson, ed., *Journals of the Colonial Legislatures of the Colonies of Vancouver Island and British Columbia, 1851–1871*, vol. 5, 28–9.
20 *The British Columbian*, 14 March, 8 August 1868; *The Mainland Guardian*, 29 October 1870.
21 Hendrickson, ed., *Journals*, vol. 5, 55.
22 Ibid., 82.
23 Stamp to W.A.G. Young, Acting Colonial Secretary, 22 January 1868, BCARS, GR 1372, B 1366, f. 1643.
24 Note written on 23 January 1868 on margin of Stamp to Young, 22 January 1868, BCARS, GR 1372, B 1366, f. 1643. Emphasis in the original.
25 Hendrickson, ed., *Journals*, vol. 4, 106.
26 British Columbia, *Sessional Papers, 1894* (Victoria: Queen's Printer, 1894), 876–80; *The Daily Columbian*, 6 March 1893.
27 'Report,' Canada, Commission on Dominion Crown Timber in British Columbia, 1897, BCARS, GR 654, v. 1, f. 2, p.11.
28 Michael Bliss, *A Living Profit: Studies in the Social History of Canadian Business, 1883–1911*, 40. See also Naylor, *History of Canadian Business*, 162–78.
29 Thomas R. Cox, *Mills and Markets: A History of the Pacific Coast Lumber Industry to 1900*, 255–83.
30 John H. Cox, 'Trade Associations in the Lumber Industry of the Pacific Northwest, 1899–1914,' 298–9.
31 William G. Robbins, *Lumberjacks and Legislators: Political Economy of the U.S. Lumber Industry, 1890–1941*, 7; see also 35–55.
32 *The Daily British Columbian*, 18 December 1889.
33 Transcript, Royal Commission on Timber and Forestry, 1909, BCARS, GR 271, v. 2, f. 2, 377–8.
34 'General Managers Report, 31st December 1908,' British Columbia Mills Timber & Trading Company, Vancouver City Archives, William McNeil Papers, Add MSS 46, v. 1, f. 6.
35 Aird Flavelle interviewed by C.D. Orchard, 20 April 1957, BCARS, Flavelle Family Papers, Add MSS 666, v. 1, f. 2, p. 9; Operating Statistics: Port Moody Cedar Mill, BCARS, Aird Flavelle Papers, Add MSS 495, v. 1, f. 18, p.1.
36 Stephen Gray, 'Forest Policy and Administration in British Columbia, 1912–1928,' 123. The economic overview of the years 1898 to 1913 was developed from *Canada Lumberman and Woodworker* 25/1 (January 1905): 34; 28/3 (February 1908): 26; *The Timberman* 14/8 (June 1913): 34; *The Labour Gazette* (1901–14).

37 *The Mainland Guardian*, 10 March 1886.
38 *Vancouver News*, 5 March 1887.
39 Agreement, 3 November 1887, University of British Columbia, Special Collections Division [hereinafter UBC], Hastings Saw Mill Company Correspondence, v. 1, f. 70.
40 *The British Columbia Commercial Journal*, 1 March, 15 March, 31 May, 7 June, 14 June, 18 October, 1 November, 14 February, 2 May, 8 August 1892; 27 March 1894.
41 *Vancouver Daily World*, 23 May, 14 June, 6 October 1892; *The Daily Columbian*, 5 October 1892; 6 July 1895.
42 *The Daily Columbian*, 28 January 1893; 2 January 1894; *Lumberman & Contractor* 2/6 (June 1905): 20.
43 Cox, *Mills and Markets*, 167–8; *Vancouver Daily World*, 18 April 1893. The Hastings operation was in financial difficulty in the 1890s: Robert A.J. McDonald, *Making Vancouver: Class, Status, and Social Boundaries, 1863–1913*, 66.
44 *The Daily Columbian*, 29 November 1895.
45 Cox, *Mills and Markets*, 270–1; Robert E. Ficken, *The Forested Land: A History of Lumbering in Western Washington*, 83.
46 *The Canada Lumberman, Wood-Workers', Manufacturers', and Millers' Gazette* [hereinafter *The Canada Lumberman*] 21/3 (March 1900): 10.
47 *The Canada Lumberman* 22/2 (February 1902): 15.
48 Memorandum of Agreement, Vancouver, 14 January 1904, UBC, Council of Forest Industries Papers, v. 81, f. 8; British Columbia Lumber and Shingle Manufacturers' Association, Price Lists, 1904–13, Council of Forest Industries of British Columbia [hereinafter COFI] Papers, v. 82, f. 1.
49 Canada, House of Commons, 'Proceedings of the Select Committee Appointed for the Purpose of Inquiring into the Prices Charges for Lumber in the Provinces of Manitoba, Alberta, and Saskatchewan,' *Journals of the House of Commons*, 1906–7, 42, part 2, appendix 6 (Ottawa: King's Printer, 1907), 362.
50 Minutes, Lumber Mills Branch, British Columbia Lumber and Shingle Manufacturers' Association, 22 February 1904, UBC, COFI Papers, v. 81, f. 8; *The Canada Lumberman* 22/2 (February 1902): 9; 24/2 (February 1904): 12; 11/10 (6 April 1904): iv; 11/36 (5 October 1904): iii; *The Labour Gazette* 4 (September 1903): 208; 6 (April 1906): 1116.
51 Transcripts, Timber and Forestry Commission, 1909, BCARS, GR 271, v. 2, f. 6, 1096, 1106, 1107.
52 Joseph Collins Lawrence, 'Markets and Capital: A History of the Lumber Industry of British Columbia (1778–1952),' 60–2.
53 *The Labour Gazette* 5 (December 1904): 561–2.

54 Minutes, Lumber Mills Branch, British Columbia Lumber and Shingle Manufacturers' Association, 15 December 1904, UBC, COFI Papers, v. 81, f. 8.
55 *The Canada Lumberman* 24/3 (August 1904): 25.
56 *The Canada Lumberman* 24/3 (March 1904): 14; 13/1 (1 February 1905): vi; *The Labour Gazette* 5 (February 1905): 834.
57 Cox, *Mills and Markets*, 294–6; R. Peter Gillis and Thomas R. Roach, *Lost Initiatives: Canada's Forest Industries, Forest Policy and Forest Conservation*, 62–3.
58 Canada, House of Commons, 'Proceedings of the Select Committee Appointed for the Purpose of Inquiring into the Prices Charged for Lumber in the Provinces of Manitoba, Alberta, and Sakatchewan,' *Journals of the House of Commons*, 1906–7, vol. 42, part 2, appendix 6 (Ottawa: King's Printer, 1907), p. viii.
59 *Canada Lumberman and Woodworker* 30/13 (1 July 1910): 25; *The Labour Gazette* 11 (September 1910): 318.
60 Memorandum, n.d., BCARS, GR 271, v. 1, f. 10.
61 *Canada Lumberman and Woodworker* 28/20 (15 October 1908): 15; *The Labour Gazette* 8 (November 1907): 533; 9 (August 1908): 148.

4: Government Policies

1 For broader perspectives on European–First Nations relations in the colonization period, see Robin Fisher, *Contact and Conflict: Indian-European Relations in British Columbia, 1774–1890*, 2d ed.; and Cole Harris, *The Resettlement of British Columbia: Essays on Colonialism and Geographical Change*.
2 Jan Peterson, *The Albernis: 1860–1922*, 29.
3 William E. Banfield to Colonial Secretary, 24 October 1859; 23 February 1860; 4 March 1860; British Columbia Archives and Records Service [hereinafter BCARS], GR 1372, B 1306, f. 107.
4 Wm. Banfield to W.A.G. Young, 3 July 1860, 6 September 1860, BCARS, GR 1372, B 1306, f. 107; Gilbert Malcolm Sproat, *Scenes and Studies of Savage Life*, 2–3; Barry M. Gough, *Gunboat Frontier: British Maritime Authority and Northwest Coast Indians, 1846–90*, 113.
5 For the state of Native treaties in British Columbia in this era, see Fisher, *Contact and Conflict*, 49–72, 146–74; Paul Tennant, *Aboriginal Peoples and Politics: The Indian Land Question in British Columbia, 1849–1989*, 1–52.
6 Edward Stamp to W.A.G. Young, 13 October 1860, BCARS, GR 1372, B 1366, f. 1643.

7 Gilbert Malcolm Sproat, 'The West Coast Indians in Vancouver Island,' 245.
8 Edward Stamp to W.A.G. Young, 4 April 1861, BCARS, GR 1372, B 1366, f. 1643.
9 Gilbert Malcolm Sproat to the Colonial Secretary, 18 May 1865, BCARS, GR 1372, B 1366, f. 1638.
10 James B. Launders to the Colonial Secretary, 3 June 1865, BCARS, GR 1372, B 1343, f. 969.
11 C. Brew to the Colonial Secretary, 7 June 1865, BCARS, GR 1372, B 1433, f. 969. For observations on the Native presence in the Lower Mainland in the nineteenth century, see Harris, *The Resettlement of British Columbia*, 69–102.
12 See H.V. Nelles, *The Politics of Development: Forests, Mines & Hydro-Electric Power in Ontario, 1849–1941*, 1–9; R. Peter Gillis and Thomas R. Roach, *Lost Initiatives: Canada's Forest Industries, Forest Policy and Forest Conservation*, 1–24.
13 W. Kaye Lamb, 'Early Lumbering on Vancouver Island, Part I: 1844–1855,' 49–51.
14 Cited in ibid., 51.
15 Edward Stamp to Col. Moody, R.E., 24 December 1858, BCARS, GR 1372, B 1366, f. 1643.
16 William A.G. Young to Captain E. Stamp, 2 January 1860, BCARS, GR 1372, B 1366, f. 1643.
17 Edward Stamp to James Douglas, 30 December 1859, BCARS, GR 1372, B 1366, f. 1643.
18 W.A.G. Young to E. Stamp, 10 January 1860, BCARS, GR 1372, B 1366, f. 1643.
19 E. Stamp to W.A.G. Young, 13 January 1860, BCARS, GR 1372, B 1366, f. 1643.
20 W.Kaye Lamb, 'Early Lumbering on Vancouver Island, Part II: 1855–1866,' 100.
21 Edward Stamp to Arthur N. Birch, 17 May 1865, BCARS, GR 1372, B 1366, f. 1643; Arthur N. Birch to Edward Stamp, 20 May 1865, BCARS, Stamp's Mill Correspondence re Sawmill on Burrard Inlet, Add MSS 120, f. 1; Indenture made between Government and the British Columbia and Vancouver Island Spar, Lumber, and Sawmill Company Limited, 17 November 1865, BCARS, Add MSS 120, f. 2.
22 Cited in Kathleen Marjorie Woodward-Reynolds, 'A History of the City and District of North Vancouver,' 18–19.
23 'Return to an Address of the Legislative Assembly, for a Return of all moneys received, during the years 1878, from Leases of Timber Lands ... ,'

British Columbia, *Sessional Papers, 1879* (Victoria: Queen's Printer, 1879), 393; Robert E. Cail, *Land, Man, and the Law: The Disposal of Crown Lands in British Columbia, 1871–1913*, 96.

24 Wm. P. Sayward to Hon. Mr Birch, Colonial Secretary, 21 March 1867, BCARS, GR 1372, B 1364, f. 1570; J.W. Trutch to W.P. Sayward, 11 May 1868, BCARS, Add MSS 120, f. 1.

25 B.C. Forest Service, Original Leases, BCARS, B4699, f. 024. Mills of all sizes, of course, continued to log privately owned timber, and individual loggers took timber from public lands.

26 British Columbia, *Sessional Papers, 1876* (Victoria: Queen's Printer, 1876), 739.

27 Cail, *Land, Man, and the Law*, 100.

28 British Columbia Forest Service, Original Licences, BCARS, B4700, v. 1.

29 *The Mainland Guardian*, 13 January 1883; *The British Columbian*, 16 December 1882; 17 February 1883.

30 British Columbia, *Sessional Papers, 1879* (Victoria: Queen's Printer, 1879), 393.

31 *The British Columbian*, 24 November, 12 December 1883; *The Mainland Guardian*, 26 January 1884.

32 *The British Columbian*, 16 May, 20 May, 27 May, 30 May, 10 June, 13 June 1885; Canada, 'Return to an Order of the House of Commons: Sessional paper No. 36A,' *Sessional Papers, 1890*, vol. 23, no. 15 (Ottawa: Queen's Printer, 1890), 2–4; *Port Moody Gazette*, 31 October 1885; 16 January 1886; Allen Seager, 'The Resource Economy, 1871–1921,' 206–9.

33 Seager, 'The Resource Economy,' 208.

34 Richard A. Rajala, 'The Receding Timber Line: Forest Practice, State Regulation, and the Decline of the Cowichan Lake Timber Industry, 1880–1992,' 181.

35 *The Daily British Columbian*, 25 April, 27 December 1887.

36 A. Haslam to Chief Commissioner of Lands and Works, 20 February 1890, BCARS, GR 1440, B 2681, f. 489/90; Cail, *Land, Man, and the Law*, 93–5.

37 Agreement re Lands, Esquimalt and Nanaimo Company and John A. Humbird, 14 January 1889, University of British Columbia, Special Collections Division [hereinafter UBC], MacMillan Bloedel Records, v. 57, f. 10; Indenture, 7 August 1895, v. 57, f. 21.

38 British Columbia, *Sessional Papers, 1892* (Victoria: Queen's Printer, 1892), 344.

39 Canada, 'Return to an Order of the House of Commons, dated 19th March 1890; – For a list of Timber Limits granted by the Dominion Government since 1st March, 1885 ..., Canada, *Sessional Papers, 1890: No. 36A* (Ottawa: Queen's Printer, 1891), 2–4.

40 Thomas Donahue to Governor Douglas, 2 February 1860, BCARS, GR 1372, B 1325, f. 479.

41 Gilbert Sproat to J.D. Pemberton, 16 January 1862, BCARS, GR 1372, B 1336, f. 905; Edward Stamp to W.A.G. Young, 10 February 1862, BCARS, GR 1372, B 1366, f. 1643.

42 Gilbert Malcolm Sproat to the Colonial Secretary, 1 November 1864; 18 May 1865, BCARS, GR 1372, B 1366, f. 1638; Matthew T. Johnston to B.W. Pearse, 15 May 1871, BCARS, GR 1372, B 1335, f. 848.

43 S.P. Moody & Co. to J.W. Trutch, 5 January 1870, BCARS, GR 1372, B 1347, f. 1159.

44 F.G. Claudet to Pearse, 2 June 1871, BCARS, GR 1372, B 1313, f. 299.

45 *The British Columbian*, 11 July 1868; 29 July 1868.

46 Indenture between Government and Jeremiah Rogers, 1867, BCARS, Add MSS 120, f. 2; Memo by Governor Seymour (?), 29 May 1868, BCARS, Add MSS 120, f. 1; Land and Works Office to Jeremiah Rogers, 26 June 1868, BCARS, GR 1372, B 1340, f. 952; Jeremiah Rogers to Governor Seymour, 10 September 1868, BCARS, GR 1372, B 1362, f. 1533; H.M. Bell to the Attorney General, 30 November 1868, BCARS, Add MSS 120, f. 1; *The British Columbian*, 5 August, 26 August 1868; E. Stamp to the Colonial Secretary, 29 July 1865, UBC, Hastings Saw Mill Correpondence, v. 1, f. 13; Captain E. Stamp to A.N. Birch, 24 January 1867, UBC, Hastings Saw Mill Correspondence, v. 1, f. 35.

47 Alexander Smith to Lands and Works Dept., 17 January 1890, BCARS, GR 1440, B 2681, f. 213/90; Petition, n.d., GR 1440, B 2681, f. 136/90; H.L. Snowden to Surveyor General, 15 April 1890, GR 1440, B 2682, f. 1009/90; Fred T. Lazenby to Surveyor General, 9 September 1890, GR 1440, B 2683, f. 2909/90.

48 A. Haslam to Chief Commissioner of Lands and Works, 20 February 1890, BCARS, GR 1440, B 2681, f. 489/90.

49 *The Daily Columbian*, 24 March 1891.

50 *Vancouver Daily World*, 6 September, 10 September 1892.

51 Brunette Saw Mill Co. to Surveyor General, 6 August 1890, BCARS, GR 1440, B 2682, f. 2424/90; Timber Inspector to Surveyor General, 1 September 1890, BCARS, GR 1440, B 2683, f. 2808/90; H.L. DeBeck to Surveyor General, 16 September 1890, B2683, f. 2988/90; D.C. Esson to Surveyor General, 13 May 1890, BCARS, GR 1440, B 2682, f. 1329/90; Brunette Saw Mill Co. to Surveyor General, 20 May 1890, BCARS, GR 1440, B 2682, f. 1385/90; Leamy & Kyle (Commercial Sawmills) to Surveyor General, 19 May 1890, BCARS, GR 1440, B 2682, f. 1417/90.

52 *Daily British Columbian*, 11 February, 24 February 1888.

53 *Vancouver Daily World*, 12 March 1890. For a dispute over log ownership in the Comox region that went to court in 1886, see Richard Somerset Mackie, *The Wilderness Profound: Victorian Life on the Gulf of Georgia*, 183–4. Concerns over private property did not disappear. In 1910, for example, holders of Dominion timber in the Fraser Valley were concerned about the security of their access to timber. The federal government favoured farmers, and the lumber interests wanted well-forested land to be reserved for logging. They also demanded that squatters be removed from their timber berths. In another instance, in 1912, after much lobbying, British Columbia lumbermen successfully forced changes to the criminal code, stiffening penalties for stealing logs, shingle bolts, and other products from the forest: Correspondence File, Vancouver City Archives, Wm. McNeil Papers, Add MSS 46, v. 2, f. 7; Wm. McNeil to Frank Oliver, 25 November 1910, v. 2, f. 8; File on Criminal Code Changes of 1912, UBC, Council of Forest Industries Papers, v. 89, f. 3.

54 *The British Columbian*, 26 August 1868; 7 June 1882; 1 August 1883; 11 July, 18 July 1885; 5 August 1887; 16 August 1889; *The Mainland Guardian*, 17 August 1870; 17 July 1872; 14 August 1875; 7 June 1882; *Dominion Pacific Herald*, 16 August 1879; *Morning Ledger*, New Westminster, 25 July, 26 July 1891; *Vancouver Daily World*, 16 August 1895; John Vye Parminter, 'An Historical Review of Forest Fire Management in British Columbia,' 7–57.

55 *Dominion Pacific Herald*, 16 August 1879; *The Daily British Columbian*, 16 August 1889; 22 February 1890; 17 August 1895.

56 *Vancouver Daily World*, 8 October, 27 October 1892.

57 Charles Melburn to Editor, *Vancouver Daily World*, 7 December 1893.

58 *Vancouver Daily World*, 18 September 1894; *The Daily Columbian*, 31 July 1895; British Columbia, 'Crown Lands Survey: Report for 1895,' *Sessional Papers, 1896* (Victoria: Queen's Printer, 1896), 734–5; British Columbia, 'Return to the House dealing with the Log Scaling Question, 15 March 1898,' *Sessional Papers, 1898* (Victoria: Queen's Printer, 1898), 809–10.

59 Transcript, Royal Commission on Timber and Forestry, 1909, BCARS, GR 271, v. 2, f. 2, p. 376.

60 *The Canada Lumberman, Wood-Workers', Manufacturers', and Millers' Gazette* 22/6 (June 1902): 13; *Canada Lumberman and Woodworker* 27/2 (February 1907): 42; 31/10 (15 May 1911): 36; *The Timberman* 12/9 (July 1911): 63; Testimony of John O'Brien, Transcript, Royal Commission on Timber and Forestry, 1909, BCARS, GR 271, v. 2, f. 2, 344–8.

61 *Vancouver Daily World*, 16 February 1894.

62 *The Daily Columbian*, 14 June, 23 July 1895.

63 'Report,' Canada, Commission on Crown Timber in British Columbia, 1897, BCARS, GR 654, v. 1, f. 2, 21.

5: Critics

1 R.E. Gosnell, *The Year Book of British Columbia and Manual of Provincial Information*, 236–7.
2 British Columbia, Department of Agriculture Report, *Sessional Papers, 1902* (Victoria: Queen's Printer, 1903), 109–21.
3 R. Peter Gillis and Thomas R. Roach, *Lost Initiatives: Canada's Forest Industries, Forest Policy and Forest Conservation*, 31–49.
4 Samuel P. Hays, *Conservation and the Gospel of Efficiency: The Progressive Conservation Movement, 1890–1920*, 2.
5 Cited in Andrew Denny Rodgers III, *Bernhard Eduard Fernow: A Story of North American Forestry*, 197.
6 J.R. Anderson, 'MS. of Memoirs,' James Robert Anderson Papers, British Columbia Archives and Records Service [hereinafter BCARS], Add MSS 1912, v. 8, f. 8, 250.
7 J.R. Anderson, 'The Preservation of Our Forests,' in British Columbia, *Sessional Papers, 1903–04* (Victoria: King's Printer, 1904), A224–A225.
8 Hays, *Conservation and the Gospel of Efficiency*, 3.
9 *The British Columbian*, New Westminster, 11 April 1865.
10 James Hett to Hon. J.W. Trutch, 11 March 1884, BCARS, British Columbia, Lands Branch Correspondence, 1872–1918, GR 1440, B 2682, f. 986/90.
11 John Edgar Gibbard, 'Early History of the Fraser Valley, 1808–1885,' 293.
12 Peter J. Smith, 'The Ideological Origins of Confederation,' 12; Janet Ajzenstat and Peter J. Smith, 'Liberal-Republicanism: The Revisionist Picture of Canada's Founding.'
13 Mark Leier, *Red Flags and Red Tape: The Making of a Labour Bureaucracy*, 93; Mark Leier, 'Rethinking Vancouver's Labour Movement: Ethnicity, Urbanism, and the Labour Aristocracy.' For an introduction to populism, see Trevor Harrison, *Of Passionate Intensity: Right-Wing Populism and the Reform Party of Canada*, 1–25. Robert A.J. McDonald, *Making Vancouver: Class, Status, and Social Boundaries, 1863–1913*, 70–9, discusses this populist/republican/labourist/radical liberal mix in late-nineteenth-century Vancouver politics.
14 *Daily Columbian*, 3 April 1893.
15 *The Mainland Guardian*, 15 April 1874.
16 *The Mainland Guardian*, 24 September 1884.

17 *The Mainland Guardian*, 15 April 1874.
18 *The Mainland Guardian*, 1 March 1876.
19 *The Mainland Guardian*, 7 March 1888.
20 *The Mainland Guardian*, 3 January 1872.
21 *The Mainland Guardian*, 1 March 1876.
22 *The Mainland Guardian*, 19 December 1885.
23 *News-Advertiser*, Vancouver, 19 June 1889.
24 W. Kaye Lamb, 'Early Lumbering on Vancouver Island, Part II: 1855–1866,' 110n.
25 *The Mainland Guardian*, 8 April 1874.
26 *The British Columbian*, 8 August 1868.
27 *The Mainland Guardian*, 8 April 1874; 2 April, 16 July 1884.
28 *Dominion Pacific Herald*, New Westminster, 22 December 1880.
29 *Vancouver Daily World*, 19 September 1890.
30 *The Mainland Guardian*, 11 May 1870.
31 *The Mainland Guardian*, 1 March 1876; 16 July 1884.
32 *The Mainland Guardian* 8 April 1874.
33 *The Mainland Guardian*, 11 November 1874.
34 *Vancouver Daily World*, 19 September 1890.
35 *The Mainland Guardian*, 10 January 1872; 6 April 1874.
36 *The Mainland Guardian*, 26 January, 27 February 1884.
37 *The Mainland Guardian*, 13 May 1874.
38 *The Mainland Guardian*, 26 February 1874.
39 *Dominion Pacific Herald*, 26 February 1879; *The British Columbian*, 16 January 1886; 14 January, 24 July 1889; 9 February 1895; *The Mainland Guardian*, 30 July 1884; 16 January 1889; F.W. Howay, *British Columbia from the Earliest Times to the Present: Biographical*, vol. 4, 724; Phyllis Veazey, 'John Hendry and the Vancouver, Westminster and Yukon Railway: "It Would Put Us on Easy Street."'
40 For a portrait of business–government collusion in this time period, see Martin Robin, *The Rush for Spoils: The Company Province, 1871–1933*, 49–86.
41 There is some evidence that British Columbians were increasingly concerned about sawmill waste and the condition of the water. In 1892 an action was launched against the Moodyville Land & Sawmill Company for polluting waters of Burrard Inlet. The company was dumping their sawdust into the Inlet: *Vancouver Daily World*, 12 September, 24 October 1892. On the other hand, sawdust floating on the water continued to be seen as a sign of prosperity: 'A yellow streak of sawdust drifting past [New]

Westminster on the Fraser river tide is ample proof that the old Ross McLaren [*sic*] mills are running again': *Lumberman & Contractor* 2/10 (October 1905): 9. For an Ontario perspective, see R. Peter Gillis, 'Rivers of Sawdust: The Battle over Industrial Pollution in Canada, 1865–1903.'

42 *Vancouver Daily World*, 15 February 1894; *Victoria Daily Colonist*, 10 February 1894.

6: Independent Logging Companies/Timber-Tenure Troubles

1 Ralph W. Hidy, Frank Ernest Hill, and Allan Nevins, *Timber and Men: The Weyerhaeuser Story*, 212; Joseph Collins Lawrence, 'Markets and Capital: A History of the Lumber Industry of British Columbia (1778–1952),' chap. 2.

2 Cited in Ken Bernsohn, *Cutting Up the North: The History of the Forest Industry in the Northern Interior*, 18–19.

3 Thomas R. Roach, 'Stewards of the Peoples' Wealth: The Founding of British Columbia's Forest Branch,' 14.

4 Ibid., 18.

5 Martin Robin, *The Rush for Spoils: The Company Province, 1871–1933*, 92–3. Peter R. Gillis and Thomas R. Roach echo this perspective in *Lost Intitiatives: Canada's Forest Industries, Forest Policy and Forest Conservation*, 140–1.

6 *The Mainland Guardian*, 12 July 1884; *The British Columbian*, 5 July 1884.

7 In 1890, for example, Royal City Planing Mills secured handlogger licences from the provincial government for Henry Long and Charles M. Nelson: Royal City Planing Mills to Chief Commissioner of Lands & Works, 10 April 1890, British Columbia Archives and Records Service [hereinafter BCARS], GR 1440, B 2682, f. 962/90.

8 Jno. B. Henderson, Dominion Saw Mill Co. to James E. Murne, 18 November 1884, BCARS, Letterbook of Correspondence Outward for the Dominion Sawmill Co., 1884–5, GR 1827, v. 2, f. 2.

9 *The Mainland Guardian*, 25 October 1879; 12 April 1873.

10 *The Mainland Guardian*, 24 November 1880; 19 March 1881; 18 March 1882; *The British Columbian*, 5 July 1882.

11 The rise in logging after 1898 was overwhelmingly on provincial land, and Dominion timber became proportionally less significant: Allen Seager, 'The Resource Economy, 1871–1921,' 221.

12 Original Timber Licences, BCARS, B4700, v. 1–3.

13 W. Innes Paterson and Thos. Magneson (B.C. Loggers Association) to the Chief Commissioner of Lands & Works, 6 February 1908, BCARS, Timber and Forestry Commission, 1909, GR 271, v. 1, f. 11.

14 Transcripts, Royal Commission on Timber and Forestry, 1909, GR 271, v. 2, f. 7, 1329.

15 G.V. Storkey, Alberta Lumber Company, to Peter Graham, Broughton Island, 8 February 1906; Storkey to Mr Smith, Coll Creek, near Port Harvey, 12 February 1906; Alberta Lumber Company to W.N. Shaw, Gabriola Island, 18 February 1907; Alberta Lumber Company to John Cox, Gabriola Island, 18 February 1907; Alberta Lumber Company to Messrs Nicholson & Gray, Port Harvey, 6 July 1906; C. McRae to Chester Homestead, Vancouver, 23 August 1907, BCARS, Alberta Lumber Company Papers, Add MSS 1293, v. 1, f. 1; v. 1, f. 2.

16 Notes on the Early History of Flavelle Cedar, 28 June 1961, BCARS, Aird Flavelle Papers, Add MSS 495, v. 1, f. 11., 13.

17 Report of the Timber Office for July 1911, Timber Inspector's Reports, 1911–12, BCARS, GR 949, v. 1, f. 10.

18 Canada, House of Commons, 'Proceedings of the Select Committee Appointed for the Purpose of Inquiring into the Prices Charged for Lumber in the Provinces of Manitoba, Alberta, and Saskatchewan,' *Journals of the House of Commons*, 1906–7, 42, part 2, appendix 6 (Ottawa: King's Printer, 1907), 378.

19 *Canada Lumberman and Woodworker* 27/2 (February 1907): 42.

20 Compilation of Abstracts from Evidence and Printed Sources, Royal Commission on Timber and Forestry, 1909, BCARS, GR 271, v. 1, f. 8.

21 Timber Inspector's Report for Year 1911, Timber Inspector's Reports, 1911–12, BCARS, GR 949, v. 1, f. 1. For production costs at the Shawnigan Lake Lumber Company from 1894 to 1915, see Robert Brian Griffin, 'The Shawnigan Lake Lumber Company, 1889–1943,' 127–8.

22 H.V. Nelles, *The Politics of Development: Forests, Mines & Hydro-Electric Power in Ontario, 1849–1941*, 71–87.

23 *The Canada Lumberman, Wood-Workers', Manufacturers', and Millers' Gazette* [hereinafter *The Canada Lumberman*] 21/6 (June 1901): 10.

24 *The Canada Lumberman*, 22/3 (March 1902): 9.

25 Thomas B. Tait to Chief Commissioner of Lands & Works, 7 December 1901, BCARS, GR 1440, B 2733, f. 7714/01; *The Canada Lumberman* 21/6 (June 1901): 10.

26 *The Canada Lumberman* 22/3 (March 1902): 9; 22/9 (September 1902): 10–11.

27 Gillis and Roach, *Lost Initiatives*, 137–9.

28 *Canada Lumberman and Woodworker* 13/1 (1 February 1905): vi; 25/12 (December 1905): 18; 26/12 (December 1906): 21; Emerson Case Papers, BCARS, GR 1208.

29 *Canada Lumberman and Woodworker* 28/8 (15 April 1908): 17; 28/18 (15 September 1908): 16; *Western Canada Lumberman* 5/9 (September 1908): 17; Joseph Austin Sayward Log Inventory, 1905–10, BCARS, K.R. Genn Collection, Add MSS 1950, v. 197, 142–6.

30 *Western Lumberman* 7/11 (November 1910): 18; Presentation of B.C. Loggers' Association, May 1910, Transcript of Hearings, Royal Commission on Timber and Forestry, 1909, BCARS, GR 271, v. 1, f. 20, 7–43; *Canada Lumberman and Woodworker* 30/9 (1 May 1910): 47.

31 *The Timberman* 12/7 (May 1911): 23 and 35.

32 Original Licences, BCARS, B4700, v. 1–3, 1899 and 1903.

33 Petition of Lumbermen and Loggers, British Columbia, *Journals of the Legislative Assembly, 1901* (Victoria: King's Printer, 1901), lvii.

34 Ibid.

35 Robert E. Cail, *Land, Man, and the Law: The Disposal of Crown Lands in British Columbia, 1971–1913*, 102.

36 The government retained the right to fix rental and royalty payments at the renewal of the option each year, and thus the government would theoretically benefit from rising timber prices. The government stopped issuing twenty-one-year timber leases. The yearly fees for Special Timber Licences were $140 on the coast, and $115 in the interior, where there was less timber per acre: *Canada Lumberman and Woodworkers* 25/4 (April 1905): 18; 25/5 (May 1905): 23; *Cail, Land, Man, and the Law*, 99, 100–3.

37 Draft of Report, 15 November 1910, Royal Commission on Timber and Forestry, 1909, BCARS, GR 271, v. 1, f. 19, 36; British Columbia, *Sessional Papers* (1906–10).

38 Transcript, 23 August 1909, BCARS, GR 271, v. 2, f. 2, 262.

39 H.N. Whitford and Roland D. Craig, *Forests of British Columbia*, 118. Licences were the most important and fastest-rising part of forest revenues. Forest revenue as a whole, including licences, leases, royalty, etc., rose from 7.2 per cent of total provincial revenue in 1901 to 15.4 per cent in 1904, to 29.4 per cent in 1907: Whitford and Craig, *Forests of British Columbia*. 120.

40 Transcript, BCARS, GR 271, v. 2, f. 6, 1117–19.

41 Agreement dated 12 January 1907, BCARS, Add MSS 495, v. 1, f. 3; Aird Flavelle interviewed by C.D. Orchard, 20 April 1957, BCARS, Flavelle Family Papers, Add MSS 666, v. 1, f. 2.

42 Aird Flavelle to Father, 2 April 1907, BCARS, Add MSS 495, v. 1, f. 2; Ken Drushka, *HR: A Biography of H.R. MacMillan*, 46–9.

43 M. Allerdale Grainger, *Woodsmen of the West*, 35.

44 Robert A.J. McDonald, 'Victoria, Vancouver, and the Economic Development of British Columbia, 1886–1914,' 386–7.

45 Donald MacKay, *Empire of Wood: The MacMillan Bloedel Story*, 66–71.
46 Notes on the Early History of Flavelle Cedar, 28 June 1961, BCARS, Add MSS 495, v. 1, f. 11, 4.
47 *The Canada Lumberman* 21/7 (July 1902): 21.
48 Aird Flavelle interviewed by C.D. Orchard, 20 April 1957, BCARS, Flavelle Family Papers, Add MSS 666, v. 1, f. 2, 6–7.
49 *Canada Lumberman and Woodworker* 27/14 (15 December 1907): 17.
50 Transcript, Forestry Commission, 1909, GR 271, v. 2, f. 2, p. 246. The importance of security of timber tenure for financing companies cannot be overestimated. In 1910 the federal government changed regulations regarding timber berths in the Railway Belt. Coastal lumbermen reacted with speed and alarm. Without security of tenure, it was difficult to get loans or debentures: 'Instances can be given where loans have been contracted in England with these licences as security for 25 year debentures, and a sinking fund created for their redemption by a fixed amount per 1000 feet of timber cut each year.' John Hendry was blunt in a telegram to Sir Wilfrid Laurier: 'Security of tenure absolutely necessary.' John Hendry, President, B.C. Lumber and Shingle Manufacturers Ass'n, to Sir Wilfrid Laurier, 27 November 1910, BCARS, Laurier Papers, Add MSS 1414, vols. 649–54, p. 177486, Reel A-183; Coast Lumbermen to Sir Wilfrid Laurier, February 1911, BCARS, Laurier Papers, vols. 665–8, pp. 181397–1452, Reel A-187.
51 Roach, 'Stewards of the People's Wealth,' 16.
52 *Canada Lumberman and Woodworker* 27/13 (1 December 1907): 14; 27/14 (15 December 1907): 17. While Continental Timber anticipated developing its timber holdings, the company got cold feet in 1908, when the coastal lumber market slackened. For his part, Clark purchased a half-interest in the Cedar Cove Lumber Company, a small mill producing lumber and shingles, in February 1909. Statement by Dr Judson Clark, Forestry Commission, 1909, BCARS, v. 1, f. 10; Transcript, Forestry Commission, 1909, BCARS, GR 271, v. 2, f. 2, pp. 373–4.
53 *Canada Lumberman and Woodworker* 28/5 (1 March 1908): 21.
54 By 1908 the independent loggers wanted any further allocation of Crown timber to be in non-transferable permits, available only to people with a proven financial commitment to log: British Columbia Loggers Association to the Chief Commissioner of Lands & Works, 6 February 1908, Minutes, British Columbia Loggers Association, 4 February 1908, University of British Columbia, Special Collections Division [hereinafter UBC], Council of Forest Industries Papers, v. 2, f. 1.
55 For information on private timber purchases gleaned from lumber-trade journals, see Lawrence, 'Markets and Capital,' 73–85. In 1910 the Victoria

Lumber & Manufacturing Company sold privately owned timberland to Andrew D. Davidson of Duluth, Minnesota, Peter Jameson of Beatrice, Nebraska; and Alexander D. McRae of Vancouver, all of whom were associated with the Canadian Western Lumber Company. The price was $3.5 million: Minutes, Meeting of Directors, St Paul, Minnesota, 3 February 1910, UBC, Humbird Family Papers, 9–1; Legal Agreements, UBC, MacMillan Bloedel Records, v. 57, f. 26. The Humbirds, owners of the Victoria Lumber & Manufacturing Company, also sold private timberland to John O'Brien of Seattle for $400,000 in 1908: Agreement for Sale, 21 September 1908, UBC, MacMillan Bloedel Records, v. 57, f. 25.

56 Existing Leases of Timberlands (31 December 1910), Draft of Report, Royal Commission on Timber and Forestry, BCARS, GR 271, v. 1, f. 19, Appendix, pp. 52–7.

57 W. Guy Flavelle, *A Cedar Saga, and the Man Who Made It Possible*, 10.

58 Transcript, Forestry Commission, 1909, BCARS, GR 271, v. 2, f. 1, 66–7; 157–71; v. 2, f. 2, 224; v. 2, f. 3, 461; v. 2, f. 6, 1027; v. 2, f. 7, 1160, 1309; Memorandum of Association of Nimpkish Lake Logging Company, BCARS, K.R. Genn Collection, Add MSS 1950, v. 110, f. 1.

59 Cail, *Land, Man, and the Law*, 99; G.W. Taylor, *Timber: History of the Forest Industry in B.C.*, 77–87.

60 Rat Portage Lbr. Co. to Deputy Chief Commissioner of Lands & Works, 21 February 1908, BCARS, GR 1440, B 2835, f. 4024/08.

61 *Canada Lumberman and Woodworker* 26/5 (May 1906): 20.

62 British Columbia, 'Timber Inspector's Report,' *Sessional Papers* (1905–9); *Canada Lumberman and Woodworker* 29/5 (1 March 1909): 19. A defence of the handlogging system, and even a call for its reinstitution, came from the labourist leaders of the Labour Councils in Vancouver and New Westminster, men who were totally against extending the term of Special Timber Licences into perpetuity. For the labourists, handlogging allowed workers with little capital to participate in an industry that was increasingly dominated by large capital. Men working cooperatively in small groups and regulated by the government could employ themselves, work small tracts of forests, curb speculation in timber, supply the mills with logs, and limit monopoly control of the resource. Handlogging, however, was doomed: Transcript, BCARS, GR 271, v. 2, f. 6, 1097 and 1101; v. 2, f. 7, 1241.

63 British Columbia, *Sessional Papers, 1906* (Victoria: King's Printer, 1906), K62–K63; *Sessional Papers, 1908* (Victoria: King's Printer, 1908), N78–N79.

64 Report of Forest Fire Fighting, Province of British Columbia, Season 1910, BCARS, GR 1125, v. 1, f. 1.

65 Whitford and Craig, *Forests of British Columbia*, 182.

66 Haslam to Fred J. Fulton, 31 July 1909, BCARS, GR 271, v. 1, f. 2; Timber Inspector Report for April 1912, BCARS, GR 949, v. 1, f. 10.

67 *The Timberman* 12/3 (January 1911): 49. Commenting on the experience in the United States, William G. Robbins, *Lumberjacks and Legislators: Political Economy of the U.S. Lumber Industry, 1890–1941,* notes that 'the evidence is overwhelming that leaders in the lumber industry used conservation propaganda as a tool to achieve stability' (p. 10). In British Columbia, too, logging operators, millmen, politicians, pulp-and-paper interests, and professional foresters used conservation language to serve their own economic interests.

68 Stephen Gray, 'The Government's Timber Business: Forest Policy and Administration in British Columbia, 1912–1928.' In 1914, after much debate, a new Royalty Act was put in place. It was intended to ensure that the provincial government received some of the benefits of rising lumber prices. However, the legislation was flawed: 'the scheme failed because it tied royalty to market prices without reference to changes in logging and milling costs. Thus when both prices and costs rose following the First World War, the royalty rates increased substantially, even though the spread between revenues and costs had not': Peter H. Pearse, 'Evolution of Forest Tenure Policy,' A8. See, too, Stephen Gray, 'Forest Policy and Administration in British Columbia, 1912–1928,' 17–51.

69 Richard Rajala, *The Legacy & the Challenge: A Century of the Forest Industry at Cowichan Lake,* 54–83; Jeremy Wilson, 'Forest Conservation in British Columbia, 1935–85: Reflections on a Barren Political Debate.'

70 The government, of course, continued to tinker with timber-tenure policies. The new Forest Branch announced in 1912 that its main tasks were fire protection, better timber-tenure regulations, and more revenue for government: cited in Gillis and Roach, *Lost Initiatives,* 149. These goals hardly suggest a radical new direction for government forest policy.

71 Mackay, *Empire of Wood,* 164–86. 'MacMillan & Bloedel Limited now [1951] operated six sawmills, two sulphate pulp mills, two plywood mills, two shingle mills, and a plant that made fuel logs from cedar waste. It controlled 747,000 acres of timber land and carried on a dozen logging operations over 175 miles of logging railway and 386 miles of logging roads. It ran the Canadian Transport Company and other subsidiaries and managed British Columbia Forest Products.

... [I]t produced 25 per cent of the 2.5 billion feet of lumber manufactured yearly on the B.C. coast. Including BCFP input, M&B marketed 32 percent. The company produced 38 percent of the coast's market pulp': Mackay, *Empire of Wood,* 174. For more on corporate concentration, see

Peter G. Aylen, 'Sustained Yield Forestry Policy in B.C. to 1956: A Deterministic Analysis of Development.' An account of the creation of British Columbia Forest Products in 1946 by financier E.P. Taylor is told in Sue Baptie, ed., *First Growth: The Story of British Columbia Forest Products Limited.*

7: Industrialization

1 Ken Drushka, *Working in the Woods: A History of Logging on the West Coast;* Ken Drushka and Hannu Konttinen, *Tracks in the Forest: The Evolution of Logging Machinery;* Wilmer Gold, *Logging as It Was: A Pictorial History of Logging on Vancouver Island;* Ed Gould, *British Columbia's Logging History;* Robert D. Turner, *Logging by Rail: The British Columbia Story.*

2 Recent scholarly work by Richard A. Rajala goes beyond this. Employing Marxist analysis and drawing on the work of Harry Braverman, he examines the logging labour process in relation to class, state, environmental, and scientific frameworks. Focusing on technology, Rajala explains changing logging practices from 1880 to 1965 in terms of the dynamic of exploitation that drives capitalism: Richard A. Rajala, *Clearcutting the Pacific Rain Forest: Production, Science, and Regulation.*

This and future chapters are more sensitive to product and labour markets in shaping the history of technology and the larger production process, but, like Rajala, emphasize the increasing industrial nature of work in the woods and the largely unquestioned dominance of capital in changing the labour process.

3 Craig Heron, 'Factory Workers,' 481. This essay provides a good introduction to debates surrounding the process of industrialization.

4 'Adventures in Vancouver Island,' and 'Diary of a Sawyer,' British Columbia Archives and Records Service [hereinafter BCARS], Pidcock Family Papers, Add MSS 728; Richard Somerset Mackie, *The Wilderness Profound: Victorian Life on the Gulf of Georgia*, 174–86.

5 Reginald Heber Pidcock, 'Diary of a Sawyer,' 8 March 1882, BCARS, Pidcock Family Papers, Add MSS 728, v. 1.

6 *The Mainland Guardian*, 6 September 1873; George Ward DeBeck, 'Reminiscences,' n.d., BCARS.

7 *The Mainland Guardian*, 14 June, 6 September 1873.

8 *The Mainland Guardian*, 28 July, 16 June 1875.

9 *The Mainland Guardian*, 4 October 1884.

10 *The British Columbian*, 18 August, 5 September 1883; *The Mainland Guardian*, 12 July 1884.

11 *The British Columbian*, 9 May, 1 September 1883; *Port Moody Gazette*, 14 June 1884; 29 August 1885; *The Mainland Guardian*, 7 October 1885.
12 *The British Columbian*, 6 May, 31 May 1882.
13 *The Daily British Columbian*, 11 September 1889.
14 *The Daily Columbian*, 6 January 1891; 4 February 1892; *Vancouver Daily World*, 23 March 1889.
15 *The Mainland Guardian*, New Westminster, 13 October 1883; *The British Columbian*, New Westminster, 13 October 1883; 21 October 1885.
16 DeBeck, 'Reminiscences,' n.d., BCARS.
17 *The Mainland Guardian*, 23 August 1882.
18 Wm. Banfield to W.A.G. Young, 6 September 1860, BCARS, Colonial Correspondence, GR 1372, B 1306, f. 107; *The British Columbian*, 1 June 1864; 27 April 1865.
19 H.L. Langevin, *British Columbia*, 43.
20 *The Mainland Guardian*, 17 June 1874.
21 *The British Columbian*, 19 July 1882.
22 *The Mainland Guardian*, 2 September 1885.
23 *The British Columbian*, 17 September 1883. For a more technical look at logging with oxen, see Drushka, *Working in the Woods*, 28–57.
24 *The Daily British Columbian*, 7 March 1888; 18 February 1892; 4 May 1894; *Vancouver Daily World*, 24 July 1890; 31 January 1891; Robert D. Turner, *Logging by Rail: The British Columbia Story*, 3.
25 Turner, *Logging by Rail*, 20–1.
26 *Western Lumberman* 8/4 (April 1911): 28; Timber Inspector Report for April 1912, BCARS, GR 949, v. 1, f. 10.
27 Richard A. Rajala, 'The Forest as Factory: Technological Change and Worker Control in the West Coast Logging Industry, 1880–1930,' 84–90.
28 *The British Columbia Commercial Journal*, 24 November 1891.
29 *Vancouver Daily World*, 21 December, 14 December 1891; 30 May, 19 August 1892.
30 *Vancouver Daily World*, 4 June 1892.
31 *Vancouver Daily World*, 27 November 1893: Rajala, 'The Forest as Factory,' 84; Drushka, *Working in the Woods*, 61. According to Robert Griffin, 'The Shawnigan Lake Lumber Company, 1889–1943,' 36, the Shawnigan Lake Lumber Company began using steam donkeys in the woods in 1889.
32 *Vancouver Daily World*, 27 November 1893.
33 *Vancouver Daily World*, 14 November 1892.
34 Rajala, 'The Forest as Factory,' 86.
35 *The Daily Columbian*, 6 November 1895; 26 June 1894.

36 *The Canada Lumberman, Wood-Workers', Manufacturers' and Millers' Gazette* 23/7 (July 1903): 13.

37 *Western Lumberman* 8/4 (April 1911): 28.

38 *The Timberman* 12/7 (May 1911): 24; Rajala, 'The Forest as Factory,' 90–102; Drushka, *Working in the Woods*, 58–98.

39 Aird Flavelle interviewed by C.D. Orchard, 20 April 1957, BCARS, Flavelle Family Papers, Add MSS 666, v. 1, f. 2, 14–15.

40 Crown Grants, Licences, and Leases held by Canadian Puget Sound Lumber Company, BCARS, Western Forests Products, Ltd., Papers, Add MSS 1996, v. 6, f. 1.

41 Agreement, August 1910, University of British Columbia, Special Collections Division [hereinafter UBC], MacMillan Bloedel Records, v. 57, f. 28.

42 —— to E.J. Palmer, 7 February 1914, Victoria Lumber & Manufacturing, Annual Statements and Reports, UBC, Annual Report, 1913, p. 2.

43 W. Curtis Sampson to A.H. Stracey, 26 March 1907, BCARS, K.R. Genn Collection, Add MSS 1950, v. 110, f. 2; Nimpkish Lake Logging Company, List of Shareholders, 6 March 1909, v. 110, f. 10; Nimpkish Lake Logging Company, Timber Limits, 25 November 1908, v. 110, f. 2; Memorandum of Association of Nimpkish Lake Logging Company, v. 110, f. 1.

44 For an introduction to scientific management and Taylorism, see Harry Braverman, *Labor and Monopoly Capital: The Degradation of Work in the Twentieth Century*. Richard Rajala discusses the debates about Taylorism and Braverman's thesis, and applies a Braverman approach to the coastal logging industry, emphasizing the introduction of physical machinery, in 'The Forest as Factory.' Graham S. Lowe links cost-accounting techniques, which reached their modern form in the first decade of the century, with Taylorism in the broad scientific-management movement around the turn of the century, where efficiency and managerial control were the main goals. See Graham S. Lowe, 'Mechanization, Feminization, and Managerial Control in the Early Twentieth-Century Canadian Office,' 178–80.

45 *The Labour Gazette* 9 (September 1908): 262.

46 *The Mainland Guardian*, 20 October 1869; *The British Columbian*, 7 February 1883.

47 W. Kaye Lamb, 'Early Lumbering on Vancouver Island, Part II: 1855–1866,' 95, 102.

48 Wm. P. Sayward to Hon. William B. Piece [*sic*], 7 October 1865, BCARS, GR 1372, B 1364, f. 1570; S.P. Moody & Co. to Chief Commissioner of Lands and Works, 27 May 1867, BCARS, GR 1372, B 1347, f. 1159.

49 *The Mainland Guardian*, 16 March 1878; *The British Columbian*, 29 March 1884; 23 September 1882; 14 November 1883.
50 Cited in Kathleen Marjorie Woodward-Reynolds, 'A History of the City and District of North Vancouver,' 29.
51 Douglas Cole and Bradley Lockner, eds., *The Journals of George M. Dawson: British Columbia, 1875–1878*, vol. 1: *1875–1876*, 116.
52 George Ward DeBeck to Keary, 16 July 1931, BCARS, 'Correspondence re Brunette Sawmill.'
53 *The British Columbian*, 29 March 1884; 8 July 1882.
54 *The British Columbian*, 23 September 1882.
55 *The British Columbian*, 14 November 1883.
56 *The Mainland Guardian*, 30 August 1884. For a Maritime perspective on the changes in carpentry and woodworking in the nineteenth century, see Ian McKay, *The Craft Transformed: An Essay on The Carpenters of Halifax, 1885–1985*, 1–26.
57 *The Daily Columbian*, 17 September, 12 May 1894; 27 April 1892; *Vancouver Daily World*, 24 July 1890; 9 May 1892; *The Truth*, New Westminster, 21 August 1890. For a discussion of saw technology in the late nineteenth century, see Thomas R. Cox, *Mills and Markets: A History of the Pacific Coast Lumber Industry to 1900*, 233–7.
58 *Canada Lumberman and Woodworker* 25/1 (January 1905): 35.
59 *The Labour Gazette* 7 (March 1907): 983–4.
60 *Canada Lumberman and Woodworker* 27/2 (February 1907): 64–7.
61 *Western Lumberman* 7/9 (September 1910): 22.
62 *The British Columbian*, 8 February 1882.
63 *Canada Lumberman and Woodworker* 27/13 (1 December 1907): 15; 30/21 (1 November 1910): 26–9; 31/10 (15 May 1911): 24–6. For changes in shingle-mill technology, see Robert Griffin, 'The Shingle Sawing Machine in British Columbia, 1901 to 1925.'
64 Richard A. Rajala, *Clearcutting the Pacific Rain Forest: Production, Science, and Regulation*, 217.

8: Loggers

1 There is an extensive literature on British Columbia labour history but only one overview: Paul A. Phillips, *No Power Greater: A Century of Labour in British Columbia*, 1–65.
2 Robert A.J. McDonald, *Making Vancouver: Class, Status, and Social Boundaries, 1863–1913*, 233. See, too, Gillian Creese, 'Class, Ethnicity, and Conflict: The Case of Chinese and Japanese Immigrants, 1880–1923.'
3 *The British Columbian*, 29 October 1864; 13 May 1882.

4 *The Labour Gazette* 4 (April 1904): 1013–14.
5 *The Timberman* 12/9 (July 1911): 61.
6 *Western Canada Lumberman* 5/4 (April 1908): 12.
7 Tabulations based on the Manuscript Census of Canada, 1881, British Columbia Archives and Records Service [hereinafter BCARS], GR 469, B 389; and the Manuscript Census of Canada, 1891, BCARS, GR 288, B 7040 and B 7041,
8 G.M. Sproat to A.C. Elliott, 7 December 1876, BCARS, GR 494, B 1101, v. 1, f. 686/76.
9 Ibid.; *The Mainland Guardian*, 16 June 1875.
10 For a discussion of Native peoples in industrial nineteenth-century British Columbia, see John Lutz, 'After the Fur Trade: The Aboriginal Labouring Class of British Columbia, 1849–1890.'
11 British Columbia Forest Service, Original Licences, BCARS, B4700, v. 1.
12 *The Daily Columbian*, 26 September 1894. See also Lutz, 'After the Fur Trade.'
13 *The Daily British Columbian*, 25 June 1888; 22 February 1887; 21 October 1889; *Vancouver Daily World*, 9 May 1891; *The Morning Ledger*, New Westminster, 20 May 1891.
14 Robert A.J. McDonald, *Making Vancouver: Class, Status, and Social Boundaries, 1863–1913*, 213.
15 Canada, 'Report of the Commissioners Appointed to Inquire into the Subject of Chinese and Japanese Immigration into the Province of British Columbia,' *Sessional Papers 1902*, no. 36, vol. 13, Sessional Paper 54 (Ottawa: King's Printer, 1902) [hereinafter Canada, Commission on Chinese and Japanese Immigration, 1902], p. 100.
16 Ibid., p. 367.
17 Timber Inspector's Report for June 1912, BCARS, GR 949, v. 1, f. 4; *The Timberman* 12/7 (May 1911): 35; Patricia E. Roy, *A White Man's Province: British Columbia Politicians and Chinese and Japanese Immigrants, 1858–1914*, 251–3.
18 *Western Lumberman* 6/9 (September 1909): 11–12; *The Canada Lumberman, Woodworkers', Manufacturers', and Millers' Gazette* 21/3 (March 1900): 10.
19 *The Mainland Guardian*, New Westminster, 14 September 1887.
20 Canada, Commission on Chinese and Japanese Immigration, 1902, 371.
21 *The Daily Herald*, Vancouver, 4 April 1888.
22 *The Daily Herald*, 23 April, 30 April, 4 May 1888; *The Daily British Columbian*, 9 May 1890; 23 February 1895; *Vancouver Daily World*, 1 December, 8 August, 10 August, 17 August, 1 October 1892.
23 See, for example, Joy Parr, *The Gender of Breadwinners: Women, Men, and*

Change in Two Industrial Towns; Mark Rosenfeld, '"It Was a Hard Life": Class and Gender in the Work and Family Rhythms of a Railway Town, 1920–1950.'

24 *The British Columbian*, 19 June, 22 June, 20 July, 24 July 1867.

25 Andrew Mason Prouty, *More Deadly than War! Pacific Coast Logging, 1827–1981*, 213. For tables on logging deaths in British Columbia after 1917, see pp. 210–13.

26 *Vancouver Daily World*, 5 August 1890; *The Daily British Columbian*, 24 August, 25 August 1888.

27 *The British Columbian*, 10 August 1865.

28 *The Mainland Guardian*, 29 July 1882; *The British Columbian*, 29 July 1882.

29 *Vancouver Daily World*, 10 May 1889.

30 *Vancouver Daily World*, 29 May 1890.

31 *The Daily Columbian*, 1 October 1894; *Vancouver Daily World*, 1 October 1894.

32 *The British Columbian*, 27 October 1883.

33 *Vancouver Daily World*, 10 June 1889.

34 'Sixth Annual Report of Chief Inspector of Machinery, 1907,' British Columbia, *Sessional Papers, 1908* (Victoria: King's Printer, 1908), p. K19.

35 *Vancouver Daily World*, 14 May 1892.

36 *Vancouver Daily World*, 2 June 1893.

37 *Vancouver Daily World*, 20 April 1895; *The Daily Columbian*, 22 April 1895.

38 *Vancouver Daily World*, 22 July, 27 December 1892.

39 *Western Lumberman* 7/7 (July 1910): 26; Testimony of R.H. Alexander, British Columbia Commission on Labour (1912–14), Transcript, BCARS, GR 684, v. 3, f. 3, 350 and 360; Minutes, General Meeting, British Columbia Loggers Association, 13 January 1908, University of British Columbia, Special Collections Division [hereinafter UBC], Council of Forest Industries Papers, v. 2, f. 1; Michael L. Hadley, *God's Little Ships: A History of the Columbia Coast Mission*.

40 McDonald, *Making Vancouver*, 213–15. See also Martin Allerdale Grainger, *Woodsmen of the West*, a novel that describes the logger's world in this period.

41 ? To Messrs. D.S.M. Co., 7 December 1884, BCARS, Add MSS 75, v. 1.

42 *Vancouver Daily World*, 25 October 1892.

43 *The Daily British Columbian*, 21 October 1889.

44 *The Daily Columbian*, 21 June 1890; 11 February 1891; 23 February 1892; 'Transcript of Hearings,' Canada, Commission on Dominion Crown Timber in British Columbia, 1897, BCARS, GR 654, v. 1, f. 4, pp. 92–3, 106, 108. Log-

gers dispensed rough justice at times. A cook working in a logging camp on the north shore of Burrard Inlet made an indecent assault on a worker who was ill in bed. When the crew returned to camp and learned of the event, they put a rope around the cook's head and soaked him in the Inlet before sending him to jail in Moodyville: *Vancouver Daily World*, 25 May 1892.

45 *Western Canada Lumberman* 5/4 (April 1908): 12.

46 *Western Canada Lumberman* 5/6 (June 1908): 15.

47 *Western Canada Lumberman* 7/4 (April 1910): 13; 7/7 (July 1910): 25 and 26; 7/11 (November 1910): 25; *Western Lumberman* 8/12 (December 1911): 26.

48 British Columbia Commission on Labour (1912–14), Transcript, BCARS, GR 684, v. 3, f. 3, 355.

49 *Western Lumberman* 7/12 (December 1910): 35.

50 *Western Lumberman* 7/7 (July 1910): 31.

51 *Lumberman & Contractor* 2/12 (December 1905): 15.

52 *The Western Wage-Earner* 2/4 (May 1910): 14.

53 *Canada Lumberman and Woodworker* 30/20 (15 October 1910): 20; 31/10 (15 May 1911): 46; *Western Lumberman* 7/10 (October 1910): 24; *The Timberman* 12/7 (May 1911): 35.

54 James S. Woodsworth, *Strangers within Our Gates, Or Coming Canadians*, 145.

55 Vernon H. Jensen, *Lumber and Labor*, 107–8.

56 Gordon Hak, 'British Columbia Loggers and the Lumber Workers Industrial Union, 1919–1922'; Gordon Hak, '"Line Up or Roll Up": The Lumber Workers Industrial Union in the Prince George District.' For an examination of the changing role of single, male hired hands in the prairie West, see Cecelia Danysk, '"A Bachelor's Paradise": Homesteaders, Hired Hands, and the Construction of Masculinity, 1880–1930.' An essay that emphasizes links between technological change and camp conditions is Richard A. Rajala's 'Bill and the Boss: Labor Protest, Technological Change, and the Transformation of the West Coast Logging Camp, 1890–1930.' For a focus on the issue of masculinity and unionization, see Todd McCallum, '"Not a Sex Question?" The One Big Union and the Politics of Radical Manhood.'

57 James Robert Conley, 'Class Conflict and Collective Action in the Working Class of Vancouver, British Columbia, 1900–1919,' 222.

58 *The Labour Gazette* 1 (June 1901): 552; 2 (July 1901): 64; Canada, Commission on Chinese and Japanese Immigration, 1902, 100.

59 Robert Brian Griffin, 'The Shawnigan Lake Lumber Company, 1889–1943,' 88.

60 *Lumberman & Contractor* 4/7 (July 1907): 12; Minutes, Lumber Mills Branch, B.C. Lumber & Shingle Manufacturers Limited, 21 January 1904, UBC, Council of Forest Industries Papers, v. 81, f. 8; *Canada Lumberman and Woodworker* 26/11 (November 1906): 22; *Lumberman & Contractor* 4/3 (March 1907): 18.
61 *The Labour Gazette* 8 (January 1908): 788.

9: Millworkers

1 For a broader pespective on social aspects of a lumber society, see Robert A.J. McDonald, 'Lumber Society on the Industrial Frontier: Burrard Inlet, 1863–1886.'
2 *Vancouver News*, 2 April 1887; *The Mainland Guardian*, 2 September 1882.
3 British Columbia, *Sessional Papers, 1894/1895* (Victoria: Queen's Printer, 1895), 593–4.
4 Tabulated from Manuscript Census of Canada, 1881, British Columbia Archives and Records Service [hereinafter BCARS], GR 469, B 389; Manuscript Census of Canada, 1891, BCARS, GR 288, B 7040 and 7041.
5 *Vancouver News*, 20 August 1886.
6 *The Mainland Guardian*, 15 January 1874.
7 Morley Roberts, *The Western Avernus, or Toil and Travel in Further North America*, 182.
8 Wm. Banfield to W.A.G. Young, 3 July and 6 September 1860; BCARS, GR 1372, B 1306, f. 107; Gilbert Malcolm Sproat to the Colonial Secretary, 1 November 1864, BCARS, GR 1372, B 1366, f. 1638; *The Mainland Guardian*, 24 December 1870; Gilbert Malcolm Sproat, *Scenes and Studies of Savage Life*, 32.
9 Sproat, *Scenes and Studies of Savage Life*, 52n.
10 G.M. Sproat to A.C. Elliott, 27 November 1876, B.C. Provincial Secretary, Records relating to Indian Affairs, 1876–8, BCARS, GR 494, B 1101, v. 1, f. 685/76.
11 *The British Columbian*, 5 July 1884.
12 Tabulated from Census of Canada, 1891, BCARS, GR 288, B 7040 and 7041.
13 *Western Canada Lumberman* 5/3 (March 1908): 11.
14 *The Labour Gazette* 7 (November 1906): 503; 13 (June 1913): 13. For First Nations grievances against the influx of Chinese and Japanese workers in the mills, ostensibly taking jobs from Native workers, see Canada, 'Report of the Commissioners Appointed to Inquire into the Subject of Chinese and Japanese Immigration into the Province of British Columbia,' *Sessional*

Papers, 1902, no. 36, vol. 13, Sessional Paper 54 (Ottawa: King's Printer, 1902) [hereinafter Canada, Commission on Chinese and Japanese Immigration, 1902], 346–7.

15 *The Labour Gazette* 5 (December 1904): 601.

16 Notes on the early History of Flavelle Cedar, 28 June 1961, Aird Flavelle Papers, BCARS, Add MSS 495, v. 1, f. 11, 8–9.

17 Canada, Commission on Chinese and Japanese Immigration, 1902, 226.

18 *The Labour Gazette* 6 (May 1906): 1234; 10 (August 1909): 173; 13 (January 1913): 722; 14 (June 1914): 1404. For more on the recruitment of French Canadians to work at Fraser River Sawmills, see *Canada Lumberman and Woodworker* 30/21 (1 November 1910): 29.

19 *The Labour Gazette* 8 (November 1907): 532; 8 (February 1908): 988.

20 British Columbia Commission on Labour (1912–14), Transcript, BCARS, GR 684, v. 3, f. 3, p. 569.

21 Canada, House of Commons, 'Proceedings of the Select Committee Appointed for the Purpose of Inquiring into the Prices Charged for Lumber in the Provinces of Manitoba, Alberta, and Saskatchewan,' *Journals of the House of Commons*, 1906–7, 42, Part II, Appendix 6 (Ottawa: King's Printer, 1907), 360.

22 British Columbia Commission on Labour (1912–14), Transcript, BCARS, GR 684, v. 3, f. 3, pp. 568–9.

23 British Columbia, Timber and Forestry Commission, 1909, Transcript, BCARS, GR 271, v. 2, f. 6., p. 1108.

24 British Columbia Commission on Labour (1912–14), Transcript, BCARS, GR 684, v. 1, f. 1., p. 162. The migratory nature of unskilled white workers was a common observation. See, for example, the testimony of A.L. Lewis, the Brunette Saw Mill manager, in 1909, GR 271, v. 2., f. 6, p. 1093.

25 *Western Canada Lumberman* 5/3 (March 1908): 11.

26 *Western Lumberman*, 7/5 (May 1910): 56.

27 For a thorough account of mill injuries and deaths in 1910, see 'First Annual Report of the Inspector of Factories of the Province of British Columbia for the Year 1910,' British Columbia, *Sessional Papers, 1911* (Victoria: King's Printer, 1911), I21–I26.

28 *The British Columbian*, 26 September 1866.

29 *Dominion Pacific Herald*, 31 March 1880; *The Mainland Guardian*, 18 June 1884.

30 *The Daily Columbian*, 23 November 1888.

31 *Vancouver Daily World*, 2 May, 13 May, 4 October, 8 October 1892; 17 September 1895.

32 *Port Moody Gazette*, 23 May 1885.
33 *The British Columbian*, 3 June, 5 July 1882; *The Mainland Guardian*, 3 June, 5 July 1882; *The Daily Columbian*, 15 September 1888.
34 *Daily British Columbian*, 19 July 1892.
35 *Vancouver Daily World*, 1 May 1890; 23 May 1891; 30 March 1892; 23 March 1893.
36 Audrey Kobayashi and Peter Jackson, 'Japanese Canadians and the Racialization of Labour in the British Columbia Sawmills,' 39; Robert A.J. McDonald, 'Lumber Society on the Industrial Frontier: Burrard Inlet, 1853–1886,' 95; *The British Columbian*, 3 April 1886; *Vancouver News*, 26 October 1886; *The Mainland Guardian*, 20 April 1887.
37 *The Daily British Columbian*, 9 June 1887; *The Mainland Guardian*, 11 June 1887; *Daily News-Advertiser*, Vancouver, 9 June, 10 June 1887.
38 *The Daily British Columbian*, 10 February 1888; 6 March, 7 March 1890.
39 *The Daily British Columbian*, 25 April 1890; 3 December 1894; *Vancouver Daily World*, 5 December 1891; 2 May 1892; Kobayashi and Jackson, 'Japanese Canadians,' 40–1; British Columbia, *Sessional Papers, 1894/1895* (Victoria: Queen's Printer, 1895), 593–4.
40 *Vancouver Daily World*, 4 July, 5 July, 6 July, 8 July, 9 July, 10 July, 11 July, 16 July, 17 July 1889; *The Daily British Columbian*, 11 July, 12 July, 18 July 1889; 2 April, 5 April 1890; *The Truth*, 3 April 1890.
41 *The Truth*, 3 July, 23 July, 25 July, 27 July 1890; *Vancouver Daily World*, 31 July 1890; 27 August 1894; *Morning Ledger*, New Westminster, 1 August 1891.
42 See Gillian Creese, 'Exclusion or Solidarity? Vancouver Workers Confront the "Oriental Problem."'
43 British Columbia Commission on Labour (1912–14), Transcript, BCARS, GR 684, v. 1, f. 7, p. 69.
44 James Robert Conley, 'Class Conflict and Collective Action in the Working Class of Vancouver, British Columbia, 1900–1919,' 239n.
45 *The Labour Gazette* 1 (May 1901): 516.
46 *The Labour Gazette* 3 (February 1903): 625.
47 Conley, 'Class Conflict and Collective Action,' 240.
48 *The Labour Gazette* 4 (July 1903): 34; 4 (August 1903): 126; Canada, 'Minutes of Evidence Taken Before the Royal Commission to Inquire into the Industrial Disputes in the Province of British Columbia,' *Sessional Papers, 1904*, no. 38, vol. 13, Sessional Paper 36a (Ottawa: King's Primer, 1904), pp. 804–9.
49 Conley, 'Class Conflict and Collective Action,' 240n.

50 *The Labour Gazette* 5 (February 1905): 831; 6 (June 1906): 1344; 7 (August 1906): 182–3; 8 (November 1907): 585; 9 (March 1909): 1009.
51 Conley, 'Class Conflict and Collective Action,' 234.
52 *Lumberman & Contractor* 4/6 (June 1907): 43; Conley, 'Class Conflict and Collective Action,' 241 and 241n.
53 Allen Seager, 'Workers, Class, and Industrial Conflict in New Westminster, 1900–1930,' 122; *The Labour Gazette* 10 (March 1910): 1002; *Canada Lumberman and Woodworker* 31/12 (15 June 1911): 51; Canada, Department of Labour, *Report on Labour Organization in Canada, 1913* (Ottawa: King's Printer, 1914), 126–33, 136–7, 168, 175.
54 *The Labour Gazette* 14 (September 1913): 260; 14 (October 1913): 418; 14 (December 1913): 677; 15 (September 1914): 366.
55 Seager, 'Workers, Class, and Industrial Conflict,' 123–9; Gordon Hak, 'British Columbia Loggers and the Lumber Workers Industrial Union, 1919–1922,' 78–9.

Conclusion

1 The Askew history is drawn from a collection of letters from Searle to Askew and Mrs Askew, British Columbia Archives and Records Service [hereinafter BCARS], Hulbert Family Papers, Add MSS 285.
2 Searle to Askew, 14 March 1867, BCARS, Add MSS 285, v. 1, f. 5.
3 Searle to Askew, 29 August 1870, BCARS, Add MSS 285, v. 1, f. 8.
4 Mortgage Title, 15 January 1879, University of British Columbia, Special Collections Division, MacMillan Bloedel Records, v. 57, f. 1; Conveyance, Askew to Boucherat, 26 September 1879, MacMillan Bloedel Records, v. 57, f. 3. Henry Croft and Henry Severne acquired the operation, and William Angus bought out Severne for $9,000 in 1884. In January 1889, Croft and Angus sold to Robert Dunsmuir, who agreed to turn the mill and property over to the Victoria Lumber & Manufacturing Company in the same year: Conveyance, Henry Severne to William Angus, 18 July 1884, MacMillan Bloedel Records, 18 July 1884, v. 57, f. 6; Indenture, 21 January 1889, MacMillan Bloedel Records, v. 57, f. 8.
5 Jerry Lembcke and William M. Tattam, *One Union in Wood: A Political History of the International Woodworkers of America*; Gordon Hak, 'British Columbia Loggers and the Lumber Workers Industrial Union, 1919–1922.'

Bibliography

I. Primary Unpublished Sources

British Columbia Archives and Records Service

Alberta Lumber Company Papers, Add MSS 1293
James Robert Anderson Papers, Add MSS 1912
British Columbia Commission on Labour (1912–14), GR 684
British Columbia Forest Service, Original Leases
British Columbia Provincial Secretary, Records relating to Indian Affairs, GR 494
British Columbia, Report of Forest Fire Fighting, 1910, GR 1125
British Columbia, Royal Commission on Forestry, 1944–5, Proceedings, GR 520
British Columbia, Royal Commission on Timber and Forestry, 1909, GR 271
Canada, Commission on Dominion Crown Timber in British Columbia, 1897, GR 654
Canada, Manuscript Census for Canada, 1881, GR 469
Canada, Manuscript Census for Canada, 1891, GR 288
Colonial Correspondence, GR 1372
Correspondence re Brunette Sawmill
George Ward DeBeck, 'Reminiscences'
Dominion Sawmill Co. Papers, Add MSS 75
Emerson Case Papers, GR 1208
Aird Flavelle Papers, Add MSS 495
Flavelle Family Papers, Add MSS 666
K.R. Genn Collection, Add MSS 1950
Hulbert Family Papers, Add MSS 285

Lands Branch Correspondence, GR 1440
Laurier Papers, Add MSS 1414
Letterbook of Correspondence Outward for the Dominion Sawmill Co., 1884–5, GR 1827
Pidcock Family Papers, Add MSS 728
Rithet Family Papers, Add MSS 504
Stamp's Mill Correspondence, Add MSS 120
Timber Inspector's Reports, 1911–12, GR 949
Western Forest Products, Ltd, Papers, Add MSS 1996

University of British Columbia, The Library, Special Collections Division

Peter B. Anderson Papers
Council of Forest Industries of British Columbia Papers
Hastings Saw Mill Correspondence
Humbird Family Papers
MacMillan Bloedel Ltd. Records
Victoria Lumber & Manufacturing Company, Annual Statements and Reports, 1906–50

Vancouver City Archives

British Columbia, Department of Lands and Works, Hastings Saw Mill, Add MSS 82
Hastings Saw Mill Company, Agreement of Sale, 1869, Add MSS 536
Hastings Saw Mill Company, Letterbook, 1870–4, Add MSS 27
William McNeill Papers, Add MSS 46

II. Primary Published Sources

Anderson, Alexander Caulfield. *The Dominion at the West: A Brief Description of the Province of British Columbia, Its Climate and Resources*. Victoria: Government Printer, 1872.

Cole, Douglas, and Bradley Lockner, eds. *The Journals of George M. Dawson: British Columbia, 1875–1878*, vol. 1: *1875–1876*. Vancouver: UBC Press, 1989.

Field, Fred W. *Capital Investments in Canada: Some Facts and Figures Respecting One of the Most Attractive Investment Fields in the World*, 2d ed. Montreal: Monetary Times of Canada, 1911.

Gosnell, R.E. *British Columbia. A Digest of Reliable Information Regarding Its*

Natural Resources and Industrial Possibilities. Vancouver: News-Advertiser Printing and Publishing, 1890.

– *The Year Book of British Columbia and Manual of Provincial Information*. Victoria, 1897.

Grainger, M. Allerdale. *Woodsmen of the West*. Toronto: McClelland & Stewart, 1964 [original edition 1908].

Harvey, Arthur. *A Statistical Account of British Columbia*. Ottawa: G.E. Desbarats, 1867.

Langevin, Hector. *British Columbia*. Ottawa: I.B. Taylor, 1872.

Macfie, Matthew. *Vancouver Island and British Columbia, Their History, Resources, and Prospects*. New York: Arno, 1973 [original edition 1865].

Matthews, J.S. *Early Vancouver*, 2 vols. Vancouver: Brock, Webber, 1932.

Mayne, Richard Charles. *Four Years in British Columbia and Vancouver Island: An Account of Their Forests, Rivers, Coasts, Gold Fields, and Resources for Colonisation*. London: J. Murray, 1862.

Roberts, Morley. *The Prey of the Strongest*. London: Hurst & Blackett, 1906[?].

– *The Western Avernus, or Toil and Travel in Further North America*. London: Smith, Elder, 1887.

Sinclair, Bertrand W. *Big Timber: A Story of the Northwest*. Boston: Little, Brown, 1916.

– *The Inverted Pyramid*. Toronto: Frederick D. Goodchild, 1924.

Sproat, Gilbert Malcolm. *British Columbia: Information for Emigrants*. London: Issued by the Agent-General for the Province, 1875 (?).

– *Scenes and Studies of Savage Life*. London: Smith, Elder, 1868.

– 'The West Coast Indians in Vancouver Island.' *Transactions of the Ethnological Society of London* 5 (1867): 243–54.

Whitford, H.N., and Roland D. Craig. *Forests of British Columbia*. Ottawa: Commission of Conservation, 1918.

Woodsworth, J.S. *Strangers within Our Gates, Or Coming Canadians*. Toronto: University of Toronto Press, 1972 [original edition 1909].

III. Newspapers and Trade Journals

The British Columbia Commercial Journal, Victoria, 1891–5

The British Columbian / The Daily British Columbian / The Daily Columbian, New Westminster

The Canada Lumberman, Wood-Workers', Manufacturers', and Millers' Gazette (The Canada Lumberman), Toronto, 1897–1904 (monthly and weekly editions)

Canada Lumberman and Woodworker, Toronto, 1905–14 (monthly and weekly editions from January 1905 to October 1907; semi-monthly thereafter)

Daily Colonist, Victoria
The Daily Herald, Vancouver
Daily News-Advertiser, Vancouver
Dominion Pacific Herald, New Westminster
Lumberman & Contractor / Western Lumberman / Western Canada Lumberman, Vancouver
The Mainland Guardian, New Westminster
Morning Ledger, New Westminster
Port Moody Gazette
The Resources of British Columbia, Victoria, 1883–5
The Timberman, Portland
The Truth, New Westminster
Vancouver Daily World
Vancouver News
Western Lumberman / Western Canada Lumberman / Lumberman & Contractor, Vancouver
The Western Wage-Earner, Vancouver, 1909–11

IV. Government Published Documents

British Columbia. *Journals of the Legislative Assembly*, 1872–1914.
– *Sessional Papers*, 1873–1914.
Canada. Department of Labour. *Report on Labour Organization in Canada, 1913*. Ottawa: King's Printer, 1914.
– House of Commons. 'Proceedings of the Select Committee Appointed for the Purpose of Inquiring into the Prices Charged for Lumber in the Provinces of Manitoba, Alberta, and Saskatchewan.' *Journals of the House of Commons*, 1906–7, 42, part II, appendix 6. Ottawa: King's Printer, 1907.
– *The Labour Gazette*, 1901–15.
– *Sessional Papers*, 1884–1910.
Hendrickson, James E., ed. *Journals of the Colonial Legislatures of the Colonies of Vancouver Island and British Columbia, 1851–1871*. Vols. 2–5. Victoria: Provincial Archives of B.C., 1980.

V. Secondary Sources: Books

Baptie, Sue, ed. *First Growth: The Story of British Columbia Forest Products Limited*. Vancouver: British Columbia Forest Products, 1975.
Barman, Jean. *The West beyond the West: A History of British Columbia*. Toronto: University of Toronto Press, 1991.

Barnes, Trevor J., and Roger Hayter, eds. *Troubles in the Rainforest: British Columbia's Forest Economy in Transition*, Canadian Western Geographical Series, Volume 33. Victoria: Department of Geography, University of Victoria, 1997.

Barnett, Homer G. *The Coast Salish of British Columbia*. 1955. Reprint, Westport, CT: Greenwood, 1975.

Bergren, Myrtle. *Tough Timber: The Loggers of B.C. – Their Story*. Toronto: Progress, 1967.

Bernsohn, Ken. *Cutting Up the North: The History of the Forest Industry in the Northern Interior*. North Vancouver: Harbour House Publishing, 1981.

Bliss, Michael, *A Living Profit: Studies in the Social History of Canadian Business*. Toronto: McClelland & Stewart, 1974.

Braverman, Harry. *Labor and Monopoly Capital: The Degradation of Work in the Twentieth Century*. New York: Monthly Review Press, 1974.

Brown, Nelson Courtlandt. *The American Lumber Industry, Embracing the Principal Features of the Resources, Production, Distribution, and Utilization of Lumber in the United States*. New York: John Wiley & Sons, 1923.

Cail, Robert E. *Land, Man, and the Law: The Disposal of Crown Lands in British Columbia, 1871–1913*. Vancouver: University of British Columbia Press, 1974.

Carlson, Keith Thor. *You Are Asked to Witness: The Stó:lō in Canada's Pacific Coast History*. Chilliwack: Stó:lō Heritage Trust, 1997.

Chandler, Alfred D., Jr. *Scale and Scope: The Dynamics of Industrial Capitalism*. Cambridge, MA: Belknap Press of Harvard University, 1990.

– *The Visible Hand: The Managerial Revolution in American Business*. Cambridge, MA: Belknap Press of Harvard University, 1977.

Clement, Wallace. *Hardrock Mining: Industrial Relations and Technological Changes at Inco*. Toronto: McClelland & Stewart, 1981.

Cox, Thomas R. *Mills and Markets: A History of the Pacific Coast Lumber Industry to 1900*. Seattle: University of Washington Press, 1974.

Cronon, William. *Nature's Metropolis: Chicago and the Great West*. New York: W.W. Norton, 1991.

Drache, Daniel, ed. *Staples, Markets, and Cultural Change: Selected Essays / Harold A. Innis*. Montreal and Kingston: McGill-Queen's University Press, 1995.

Drushka, Ken. *HR: A Biography of H.R. MacMillan*. Madeira Park, BC: Harbour, 1995.

– *Working in the Woods: A History of Logging on the West Coast*. Madeira Park, BC: Harbour, 1992.

Drushka, Ken, and Hannu Konttinen. *Tracks in the Forest: The Evolution of Logging Machinery*. Helsinki: Timberjack Group, 1997.

Ficken, Robert E. *The Forested Land: A History of Lumbering in Western Washington*. Seattle: University of Washington Press, 1987.

Fisher, Robin. *Contact and Conflict: Indian–European Relations in British Columbia, 1774–1890*, 2d ed. Vancouver: UBC Press, 1992.

Flavelle, W. Guy. *A Cedar Saga, and the Man Who Made It Possible* Port Moody, BC: Agency, 1966.

Galois, Robert. *Kwakwaka'wakw Settlement, 1775–1920: A Geographical Analysis and Gazetteer*. Vancouver: UBC Press, 1994.

Gibson, Gordon. *Bull of the Woods: The Gordon Gibson Story*. Vancouver: Douglas & McIntyre, 1980.

Gillis, R. Peter, and Thomas R. Roach. *Lost Initiatives: Canada's Forest Industries, Forest Policy and Forest Conservation*. Westport, CT: Greenwood, 1986.

Gold, Wilmer. *Logging as It Was: A Pictorial History of Logging on Vancouver Island*. Victoria: Morriss, 1985.

Gough, Barry M. *Gunboat Frontier: British Maritime Authority and Northwest Coast Indians, 1846–90*. Vancouver: UBC Press, 1984.

Gould, Ed. *British Columbia's Logging History*. North Vancouver: Hancock House, 1975.

Gresko, Jacqueline, and Richard Howard, eds. *Fraser Port: Freightway to the Pacific, 1858–1985*. Victoria: Sono Nis, 1986.

Gunton, Thomas I. *Resources, Regional Development and Provincial Policy: A Case Study of British Columbia*. Ottawa: Canadian Centre for Policy Alternatives, 1982.

Hadley, Michael L. *God's Little Ships: A History of the Columbia Coast Mission*. Madeira Park, BC: Harbour, 1995.

Harris, Cole. *The Resettlement of British Columbia: Essays on Colonialism and Geographical Change*. Vancouver: UBC Press, 1997.

Harrison, Trevor. *Of Passionate Intensity: Right-Wing Populism and the Reform Party of Canada*. Toronto: University of Toronto Press, 1995.

Hays, Samuel P. *Conservation and the Gospel of Efficiency: The Progressive Conservation Movement, 1890–1920*. New York: Atheneum, 1969 [original edition 1959].

Hidy, Ralph W., Frank Ernest Hill, and Allan Nevins. *Timber and Men: The Weyerhaeuser Story*. New York: Macmillan, 1963.

Howay, F.W. *British Columbia from Earliest Times to the Present*. Vancouver: S.J. Clark, 1914.

International Longshoremen's and Warehousemen's Union [ILWU] Local 500 Pensioners. *'Man Along the Shore!' The Story of the Vancouver Waterfront*. Vancouver: ILWU Local 500, 1975.

Jensen, Vernon H. *Lumber and Labor*. New York: Farrar & Rinehart, 1945.

Johnston, Hugh J.M., ed. *The Pacific Province: A History of British Columbia*. Vancouver: Douglas & McIntyre, 1996.

Kahrer, Gabrielle. *From Speculative to Spectacular ... The Seymour River Valley, 1870's to 1980's: A History of Resource Use*. Burnaby, BC: Greater Vancouver Regional District Parks Department, 1989.

Kealey, Gregory S. *Toronto Workers Respond to Industrial Capitalism, 1867–1892*. Toronto: University of Toronto Press, 1980.

Kirkland, Edward C. *Industry Comes of Age: Business, Labor, and Public Policy, 1860–1897*. New York: Holt, Rinehart & Winston, 1961.

Knight, Rolf. *Indians at Work: An Informal History of Native Labour in British Columbia, 1858–1930*. Vancouver: New Star, 1996.

Lambert, Richard S., with Paul Pross. *Renewing Nature's Wealth*. Toronto: Department of Lands and Forests, 1967.

Leier, Mark. *Red Flags and Red Tape: The Making of a Labour Bureaucracy*. Toronto: University of Toronto Press, 1995.

– *Where the Fraser River Flows: The Industrial Workers of the World in British Columbia*. Vancouver: New Star, 1990.

Lembcke, Jerry, and William M. Tattam. *One Union in Wood: A Political History of the International Woodworkers of America*. Madeira Park, BC: Harbour, 1984.

Loo, Tina. *Making Law, Order, and Authority in British Columbia, 1821–1871*. Toronto: University of Toronto Press, 1994.

MacDonald, Norbert. *Distant Neighbors: A Comparative History of Seattle & Vancouver*. Lincoln: University of Nebraska Press, 1987.

MacKay, Donald. *Empire of Wood: The MacMillan Bloedel Story*. Vancouver: Douglas & McIntyre, 1982.

– *The Lumberjacks*. Toronto: McGraw-Hill Ryerson, 1978.

Mackie, Richard Somerset. *Trading beyond the Mountains: The British Fur Trade on the Pacific*. Vancouver: UBC Press, 1997.

– *The Wilderness Profound: Victorian Life on the Gulf of Georgia*. Victoria: Sono Nis, 1995.

Marchak, Patricia. *Green Gold: The Forest Industry in British Columbia*. Vancouver: UBC Press, 1983.

McDonald, Robert A.J. *Making Vancouver: Class, Status, and Social Boundaries, 1863–1913*. Vancouver: UBC Press, 1996.

McKay, Ian. *The Craft Transformed: An Essay on the Carpenters of Halifax, 1885–1985*. Halifax: Holdfast, 1985.

Morton, James. *The Enterprising Mr. Moody, the Bumptious Captain Stamp: The Lives and Colourful Times of Vancouver's Lumber Pioneers*. Vancouver: J.J. Douglas, 1977.

Mouat, Jeremy. *Roaring Days: Rossland's Mines and the History of British Columbia*. Vancouver: UBC Press, 1995.
Nelles, H.V. *The Politics of Development: Forests, Mines and Hydro-Electric Power in Ontario, 1849–1941*. Toronto: Macmillan of Canada, 1974.
Ormsby, Margaret A. *British Columbia: A History*. Toronto: Macmillan of Canada, 1958.
Palmer, Bryan D. *Working-Class Experience: Rethinking the History of Canadian Labour, 1800–1991*, 2d ed. Toronto: McClelland & Stewart, 1992.
Parr, Joy. *The Gender of Breadwinners: Women, Men, and Change in Two Industrial Towns*. Toronto: University of Toronto Press, 1990.
Pearse, Peter H. *Timber Rights and Forest Policy in British Columbia*, Report of the Royal Commission on Forest Resources, vol. 2. Victoria, BC: Queen's Printer, 1976.
Perreault, E.G. *Wood & Water: The Story of Seaboard Lumber and Shipping*. Vancouver: Douglas & McIntyre, 1985.
Peterson, Jan. *The Albernis: 1860–1922*. Lantzville, BC: Oolichan, 1992.
Phillips, Paul. *No Power Greater: A Century of Labour in British Columbia*. Vancouver: BC Federation of Labour, 1967.
Prouty, Andrew Mason. *More Deadly Than War! Pacific Coast Logging, 1827–1981*. New York: Garland, 1985.
Rajala, Richard A. *Clearcutting the Pacific Rain Forest: Production, Science, and Regulation*. Vancouver: UBC Press, 1998.
– *The Legacy & the Challenge: A Century of the Forest Industry at Cowichan Lake*. Lake Cowichan, BC: Lake Cowichan Heritage Advisory Committee, 1993.
Reitano, Joanne. *The Tariff Question in the Gilded Age: The Great Debate of 1888*. University Park: Pennsylvania State University Press, 1994.
Robbins, William G. *Colony and Empire: The Capitalist Transformation of the American West*. Lawrence: University Press of Kansas, 1994.
Robbins, William G. *Hard Times in Paradise: Coos Bay, Oregon, 1850–1896*. Seattle: University of Washington Press, 1988.
– *Lumberjacks and Legislators: Political Economy of the U.S. Lumber Industry, 1890–1941*. College Station: Texas A&M University Press, 1982.
Robin, Martin. *The Rush for Spoils: The Company Province, 1871–1933*. Toronto: McClelland & Stewart, 1972.
Rodgers III, Andrew Denny. *Bernard Eduard Fernow: A Story of North American Forestry*. Princeton, NJ: Princeton University Press, 1951.
Roy, Patricia E. *A White Man's Province: British Columbia Politicians and Chinese and Japanese Immigrants, 1858–1914*. Vancouver: UBC Press, 1989.
Shakespeare, Mary, and Rodney H. Pain. *West Coast Logging, 1840–1910*. Mercury Series, History Division Paper No. 22. Ottawa: National Museums of Canada, 1977.

Sprague, D.N. *Post-Confederation Canada: The Structure of Canadian History since Confederation*. Scarborough: Prentice-Hall, 1990.

Taylor, C. James. *The Heritage of the British Columbia Forest Industry: A Guide for Planning, Selection and Interpretation of Sites*. Ottawa: National Historic Parks and Sites, 1987.

Taylor, Graham D., and Peter A. Baskerville. *A Concise History of Business in Canada*. Toronto: Oxford University Press, 1994.

Taylor, G.W. *Timber: The History of the Forest Industry in B.C.* Vancouver: J.J. Douglas, 1975.

Tennant, Paul. *Aboriginal Peoples and Politics: The Indian Land Question in British Columbia, 1849–1989*. Vancouver: UBC Press, 1990.

Turner, Robert D. *Logging by Rail: The British Columbia Story*. Victoria: Sono Nis, 1990.

White, Richard. *Land Use, Environment, and Social Change: The Shaping of Island County, Washington*. Seattle: University of Washington Press, 1980.

Wynn, Graeme. *Timber Colony: A Historical Geography of Early Nineteenth-Century New Brunswick*. Toronto: University of Toronto Press, 1981.

VI. Secondary Sources: Articles and Theses

Ajzenstat, Janet, and Peter J. Smith. 'Liberal-Republicanism: The Revisionist Picture of Canada's Founding.' In *Canada's Origins: Liberal, Tory, or Republican*, ed. Janet Ajzenstat and Peter J. Smith, 1–18. Ottawa: Carleton University Press, 1995.

Altmeyer, George. 'Three Ideas of Nature in Canada, 1893–1914.' *Journal of Canadian Studies* 11/3 (August 1976): 21–36.

Aylen, Peter G. 'Sustained Yield Forestry Policy in B.C. to 1956: A Deterministic Analysis of Development.' MA thesis, University of Victoria, 1984.

Belshaw, John Douglas. 'Provincial Politics, 1871–1916.' In *The Pacific Province: A History of British Columbia*, ed. Hugh J.M. Johnston, 134–64. Vancouver: Douglas & McIntyre, 1996.

Brookes, Alan A. 'Out-Migration from the Maritime Provinces, 1860–1900: Some Preliminary Considerations.' *Acadiensis* 5/2 (Spring 1976): 26–55.

Brooks, George Waite Stirling. 'Edgar Crow Baker: An Entrepreneur in Early British Columbia.' MA thesis, University of British Columbia, 1976.

– 'Edgar Crow Baker: An Entrepreneur in Early British Columbia.' *BC Studies* 31 (Autumn 1976): 23–43.

Careless, J.M.S. 'The Business Community in the Early Development of Victoria, British Columbia.' In *Canadian Business History: Selected Studies, 1497–1971*, ed. David S. Macmillan, 104–23. Toronto: McClelland & Stewart, 1972.

– 'The Lowe Brothers, 1852–70: A Study in Business Relations on the North Pacific Coast.' *BC Studies* 2 (Summer 1969): 1–18.

Carmichael, Herbert. 'Pioneer Days in Pulp and Paper.' *British Columbia Historical Quarterly* 9/3 (July 1945): 201–12.

Carrothers, W.A. 'Forest Industries of British Columbia.' In *The North American Assault on the Canadian Forest: A History of the Lumber Trade between Canada and the United States*, ed. A.R.M. Lower, 227–344. Toronto: Ryerson, 1938.

Clausen, Susanne. 'The Plywood Girls: Women and Gender Ideology at the Port Alberni Plywood Plant, 1942–1992.' *Labour/Le Travail* 41 (Spring 1998): 199–235.

Conley, James Robert. 'Class Conflict and Collective Action in the Working Class of Vancouver, British Columbia, 1900–1919.' PhD thesis, Carleton University, 1986.

Cox, John H. 'Trade Associations in the Lumber Industry of the Pacific Northwest, 1899–1914.' *The Pacific Northwest Quarterly* 41/4 (October 1950): 285–311.

Craven, Paul. 'Labour and Management on the Great Western Railway.' In *Labouring Lives: Work and Workers in Nineteenth-Century Ontario*, ed. Paul Craven, 335–410. Toronto: University of Toronto Press, 1995.

Creese, Gillian. 'Class, Ethnicity, and Conflict: The Case of Chinese and Japanese Immigrants, 1880–1923.' In *Workers, Capital, and the State in British Columbia: Selected Papers*, ed. Rennie Warburton and David Coburn, 55–85. Vancouver: UBC Press, 1988.

– 'Exclusion or Solidarity? Vancouver Workers Confront the "Oriental Problem,"' *BC Studies* 80 (Winter 1988–9): 24–51.

Danysk, Cecelia. '"A Bachelor's Paradise": Homesteaders, Hired Hands, and the Construction of Masculinity, 1880–1930.' In *Making Western Canada: Essays on European Colonization and Settlement*, ed. Catherine Cavanaugh and Jeremy Mouat, 154–85. Toronto: Garamond, 1996.

Fairholm, Olive. 'John Robson and Confederation.' In *British Columbia and Confederation*, ed. W. George Shelton, 97–123. Victoria: University of Victoria Press, 1967.

Flynn, James E. 'Early Lumbering on Burrard Inlet, 1862–1891.' Bachelor of Applied Science in Forest Engineering thesis, University of British Columbia, 1942.

Gallacher, Daniel T. 'John Muir.' *Dictionary of Canadian Biography*, vol. 11: *1881 to 1890*. Toronto: University of Toronto Press, 1982.

Gibbard, John Edgar. 'Early History of the Fraser Valley, 1808–1885.' MA thesis, University of British Columbia, 1937.

Gillis, R. Peter. 'The Ottawa Lumber Barons and the Conservation Movement, 1880–1914.' *Journal of Canadian Studies* 9 (February 1974): 14–31.

– 'Rivers of Sawdust: The Battle over Industrial Pollution in Canada, 1865–1903.' *Journal of Canadian Studies* 21/1 (Spring 1986): 84–103.

Gough, Barry M. 'The Character of the British Columbia Frontier.' *BC Studies* 32 (Winter 1976–7): 28–40.

– 'Forests and Sea Power: A Vancouver Island Economy, 1778–1885.' *Journal of Forest History* 32/3 (July 1988): 117–24.

Gray, Stephen. 'Forest Policy and Administration in British Columbia, 1912–1928.' MA thesis, Simon Fraser University, 1982.

– 'The Government's Timber Business: Forest Policy and Administration in British Columbia, 1912–1928.' *BC Studies* 81 (Spring 1989): 24–49.

Griffin, Robert Brian. 'The Shawnigan Lake Lumber Company, 1889–1943.' MA thesis, University of Victoria, 1979.

– 'The Shingle Sawing Machine in British Columbia, 1901 to 1925.' *Material History Bulletin* 13 (Fall 1981): 21–38.

Hak, Gordon. 'British Columbia Loggers and the Lumber Workers Industrial Union, 1919–1922.' *Labour/Le Travail* 23 (Spring 1989): 67–90.

– '"Line Up or Roll Up": The Lumber Workers Industrial Union in the Prince George District.' *BC Studies* 86 (Summer 1990): 57–74.

– 'On the Fringes: Capital and Labour in the Forest Economies of the Port Alberni and Prince George Districts, British Columbia, 1910–1939.' PhD thesis, Simon Fraser University, 1986.

– 'Red Wages: Communists and the 1934 Vancouver Island Loggers Strike.' *Pacific Northwest Quarterly* 80/3 (July 1989): 82–90.

Heron, Craig. 'Factory Workers.' In *Labouring Lives: Work and Workers in Nineteenth-Century Ontario*, ed. Paul Craven, 480–511. Toronto: University of Toronto Press, 1995.

Heron, Craig, and Robert Storey. 'On the Job in Canada.' In *On the Job: Confronting the Labour Process in Canada*, ed. Craig Heron and Robert Storey, 3–46. Montreal and Kingston: McGill-Queen's University Press, 1986.

– 'The Second Industrial Revolution in Canada, 1890–1930.' In *Class, Community and the Labour Movement: Wales and Canada, 1850–1930*, ed. Deian R. Hopkin and Gregory S. Kealey, 48–66. Aberytstwyth: Llafur/Canadian Committee on Labour History, 1989.

Howay, F.W. 'Early Shipping in Burrard Inlet: 1863–1870.' *British Columbia Historical Quarterly* 1/1 (January 1937): 3–20.

Kahrer, Anna Gabrielle. 'Logging and Landscape Change on the North Shore of Burrard Inlet, British Columbia, 1860s to 1930s.' MA thesis, University of British Columbia, 1988.

Kobayashi, Audrey, and Peter Jackson. 'Japanese Canadians and the Racialization of Labour in the British Columbia Sawmills.' *BC Studies* 103 (Autumn 1994): 33–58.

Lamb, W. Kaye. 'Early Lumbering on Vancouver Island, Part I: 1844–1855.' *British Columbia Historical Quarterly* 2/1 (January 1938): 31–53.

– 'Early Lumbering on Vancouver Island, Part II: 1855–1866.' *British Columbia Historical Quarterly* 2/2 (April 1938): 95–121.

Lawrence, Joseph Collins. 'Markets and Capital: A History of the Lumber Industry of British Columbia (1778–1952).' MA thesis, University of British Columbia, 1957.

Leier, Mark. 'Rethinking Vancouver's Labour Movement: Ethnicity, Urbanism, and the Labour Aristocracy.' *Canadian Historical Review* 74 (1993): 125–65.

Lowe, Graham S. 'Mechanization, Feminization, and Managerial Control in the Early Twentieth-Century Canadian Office.' In *On the Job: Confronting the Labour Process in Canada*, ed. Craig Heron and Robert Storey, 177–209. Montreal and Kingston: McGill-Queen's University Press, 1986.

Lower, A.R.M. 'The Trade in Square Timber.' In *Approaches to Canadian Economic History*, ed. W.T. Easterbrook and M.H. Watkins, 28–48. Toronto: McClelland & Stewart, 1967.

Lutz, John. 'After the Fur Trade: The Aboriginal Labouring Class of British Columbia, 1849–1890.' *Journal of the Canadian Historical Association* (1992): 69–93.

Mackie, Richard. 'Colonial Land, Indian Labour and Company Capital: The Economy of Vancouver Island, 1849–1858.' MA thesis, University of Victoria, 1984.

– 'William Parsons Sayward.' *Dictionary of Canadian Biography*, vol. 13: *1901 to 1910*. Toronto: University of Toronto Press, 1994.

Mackintosh, W.A. 'Economic Factors in Canadian History.' In *Approaches to Canadian Economic History*, ed. W.T. Easterbrook and M.H. Watkins, 1–15. Toronto: McClelland & Stewart, 1967.

Marchak, Patricia. 'A Changing Global Context for British Columbia's Forest Industry.' In *Troubles in the Rainforest: British Columbia's Forest Economy in Transition*, ed. Trevor J. Barnes and Roger Hayter, 149–64. Canadian Western Geographical Series, vol. 33. Victoria: Department of Geography, University of Victoria, 1997.

Marris, Robert Howard, '"Pretty Sleek and Fat": The Genesis of Forest Policy in British Columbia, 1903–1914.' MA thesis, University of British Columbia, 1979.

McCallum, Todd. '"Not a Sex Question?" The One Big Union and the Politics of Radical Manhood.' *Labour/Le Travail* 42 (Fall 1998): 15–54.

McDonald, Margaret Lillooet. 'New Westminster, 1859–1871.' MA thesis, University of British Columbia, 1947.

McDonald, Robert A.J. 'Business Leaders in Early Vancouver, 1886–1914.' PhD thesis, University of British Columbia, 1977.

– 'Lumber Society on the Industrial Frontier: Burrard Inlet, 1863–1886.' *Labour/Le Travail* 33 (Spring 1994): 69–96.

– 'Victoria, Vancouver, and the Economic Development of British Columbia, 1886–1914.' In *British Columbia: Historical Readings*, ed. W. Peter Ward and Robert A.J. McDonald, 369–95. Vancouver: Douglas & McIntyre, 1981.

McKay, Ian. 'Capital and Labour in the Halifax Baking and Confectionary Industry during the Last Half of the Nineteenth Century.' *Labour/Le Travailleur* 3 (1978): 63–108.

Mouat, Jeremy. 'Creating a New Staple: Capital, Technology and Monopoly in B.C.'s Resource Sector, 1901–1925.' *Journal of the Canadian Historical Association*, New ser., 1 (1990): 215–37.

Naylor, R.T. *The History of Canadian Business, 1867–1914*. Vol. 2: *Industrial Development*. Toronto: Lorimer, 1975.

Parminter, John Vye. 'An Historical Review of Forest Fire Management in British Columbia.' Master of Forestry thesis, University of British Columbia, 1978.

Phillips, Paul A. 'Confederation and the Economy of British Columbia.' In *British Columbia and Confederation*, ed. W. George Shelton, 43–66. Victoria: University of Victoria, 1967.

– 'The Underground Economy: The Mining Frontier to 1920.' In *Workers, Capital, and the State in British Columbia: Selected Papers*, ed. Rennie Warburton and David Coburn, 35–54. Vancouver: UBC Press, 1988.

Radforth, Ian. 'The Shantymen.' In *Labouring Lives: Work and Workers in Nineteenth-Century Ontario*, ed. Paul Craven, 204–78. Toronto: University of Toronto Press, 1995.

Rajala, Richard A. 'Bill and the Boss: Labor Protest, Technological Change, and the Transformation of the West Coast Logging Camp, 1890–1930.' *Journal of Forest History* 33/4 (October 1989): 168–79.

– 'Clearcutting the British Columbia Coast: Work, Environment and the State, 1880–1930.' In *Making Western Canada: Essays on European Colonization and Settlement*, ed. Catherine Cavanaugh and Jeremy Mouat, 104–32. Toronto: Garamond, 1996.

– 'The Forest as Factory: Technological Change and Worker Control in the West Coast Logging Industry, 1880–1930.' *Labour/Le Travail* 32 (Fall 1993): 73–104.

– 'The Receding Timber Line: Forest Practice, State Regulation, and the Decline of the Cowichan Lake Timber Industry, 1880–1992.' In *Canadian*

Papers in Business History, vol. 2, ed. Peter A. Baskerville, 179–209. Victoria: Public History Group, University of Victoria, 1993.
– 'The Rude Science: A Social History of West Coast Logging, 1890–1930.' MA thesis, University of Victoria, 1987.
Ralston, Keith. 'Patterns of Trade and Investment on the Pacific Coast, 1867–1892: The Case of the British Columbia Canning Industry.' In *British Columbia: Historical Readings*, ed. W. Peter Ward and Robert A.J. McDonald, 296–305. Vancouver: Douglas & McIntyre, 1981.
Reid, David J. 'Company Mergers in the Fraser River Canning Industry, 1885–1902.' In *British Columbia: Historical Readings*, ed. W. Peter Ward and Robert A.J. McDonald, 306–27. Vancouver: Douglas & McIntyre, 1981.
Reid, Richard M. 'James Maclaren.' *Dictionary of Canadian Biography*, vol. 12: *1891–1900*. Toronto: University of Toronto Press, 1990.
Roach, Thomas R. 'Stewards of the People's Wealth: The Founding of British Columbia's Forest Branch.' *Journal of Forest History* 28/1 (January 1984): 14–23.
Robertson, Irene Elaine. 'The Business Community and the Development of Victoria, 1858–1900.' MA thesis, University of Victoria, 1981.
Rosenfeld, Mark. '"It Was A Hard Life": Class and Gender in the Work and Family Rhythms of a Railway Town.' Canadian Historical Association *Historical Papers* (1988): 237–79.
Roy, Patricia E. 'A Choice between Evils: The Chinese and the Construction of the Canadian Pacific Railway in British Columbia.' In *The CPR West: The Iron Road and the Making of a Nation*, ed. Hugh A. Dempsey, 13–34. Vancouver: Douglas & McIntyre, 1984.
– 'The Fine Arts of Lobbying and Persuading: The Case of the B.C. Electric Railway, 1897–1917.' In *Canadian Business History: Selected Studies, 1497–1971*, ed. David S. Macmillan, 239–61. Toronto: McClelland and Stewart, 1972.
– 'The Preservation of Peace in Vancouver: The Aftermath of the Anti-Chinese Riot of 1887.' *BC Studies* 31 (Autumn 1976): 44–59.
Schieder, Rupert. 'Woodsman of the West: Martin Allerdale Grainger.' *Forest History* 11 (October 1967): 6–13.
Seager, Allen. 'The Resource Economy, 1871–1921.' In *The Pacific Province: A History of British Columbia*, ed. Hugh J.M. Johnston, 205–52. Vancouver: Douglas & McIntyre, 1996.
– 'Workers, Class, and Industrial Conflict in New Westminster, 1900–1930.' In *Workers, Capital, and the State in British Columbia: Selected Papers*, ed. Rennie Warburton and David Coburn, 117–40. Vancouver: UBC Press, 1988.

Smith, Peter J. 'The Ideological Origins of Confederation.' *Canadian Journal of Political Science* 20/1 (March 1987): 3–29

Tomich, Dale. 'World of Capital / Worlds of Labor: A Global Perspective.' In *Reworking Class*, ed. John R. Hall, 287–311. Ithaca, NY: Cornell University Press, 1997.

Veazey, Phyllis. 'John Hendry and the Vancouver, Westminster and Yukon Railway: "It Would Put Us on Easy Street."' *BC Studies* 59 (Autumn 1983): 44–63.

Watt, Robert D. 'Early B.C. Sawmill Machinery: 1869.' *Material History Bulletin*, 21 (1977), 47–54.

Watkins, M.H. 'A Staple Theory of Economic Growth.' In *Approaches to Canadian Economic History*, ed. W.T. Easterbrook and M.H. Watkins, 49–73. Toronto: McClelland & Stewart, 1967.

Webber, Jeremy. 'Labour and the Law.' In *Labouring Lives: Work and Workers in Nineteenth-Century Ontario*, ed. Paul Craven, 105–201. Toronto: University of Toronto Press, 1995.

Wilson, Jeremy. 'Forest Conservation in British Columbia, 1935–85: Reflections on a Barren Political Debate.' *BC Studies* 76 (Winter, 1987–8): 3–32.

'Woods Work.' *Whistle Punk* 1/1 (Spring 1984): 22–3

Woodward-Reynolds, Kathleen Marjorie. 'A History of the City and District of North Vancouver.' MA thesis, University of British Columbia, 1943.

Yarmie, Andrew. 'The Right to Manage: Vancouver Employers' Associations, 1900–1923.' *BC Studies* 90 (Summer 1991): 40–74.

Yerburgh, Richard Eustre Marryat. 'An Economic History of Forestry in British Columbia.' MA thesis, University of British Columbia, 1931.

Index